The
Workflow
Imperative

The Workflow Imperative

Thomas M. Koulopoulos

Building Real World Business Solutions

VAN NOSTRAND REINHOLD

I(T)P™ A Division of International Thomson Publishing Inc.

New York • Albany • Bonn • Boston • Detroit • London • Madrid • Melbourne
Mexico City • Paris • San Francisco • Singapore • Tokyo • Toronto

Copyright © 1995 by Thomas M. Koulopoulos

 Published by Van Nostrand Reinhold, a division of
International Thomson Publishing Inc.
The ITP logo is a trademark under license

Printed in the United States of America
For more information, contact:

Van Nostrand Reinhold
115 Fifth Avenue
New York, NY 10003

Chapman & Hall GmbH
Pappelallee 3
69469 Weinheim
Germany

Chapman & Hall
2-6 Boundary Row
London
SE1 8HN
United Kingdom

International Thomson Publishing Asia
221 Henderson Road #05-10
Henderson Building
Singapore 0315

Thomas Nelson Australia
102 Dodds Street
South Melbourne, 3205
Victoria, Australia

International Thomson Publishing Japan
Hirakawacho Kyowa Building, 3F
2-2-1 Hirakawacho
Chiyoda-ku, 102 Tokyo
Japan

Nelson Canada
1120 Birchmount Road
Scarborough, Ontario
Canada M1K 5G4

International Thomson Editores
Campos Eliseos 385, Piso 7
Col. Polanco
11560 Mexico D.F. Mexico

4 5 6 7 8 9 10 RRDHB 01 00 99 98 97

Library of Congress Cataloging-in-Publication Data
Koulopoulos, Thomas M.
    The workflow imperative : building real world business
solutions / by Thomas Koulopoulos.
        p. cm.
    Originally published: 1st ed. Boston : Delphi Pub., c1994.
    Includes index.
    ISBN 0-442-01975-0
    1. Office practice—Automation. 2. Office procedures. 3. Office
management. I. Title.
HF5548.K59 1995
651.8—dc20   94-46694
                                                CIP

Ann Finnerty • Freehand Illustrations

Management: Jo-Ann Campbell • Production: mle design, 562 Milford Point Rd, Milford, CT 06460 • 203-878-3793

To Debra–

    partner in business and in life,
    for her dedication and vision.

Contents

Preface

Over the past five years, I have been involved in analyzing the business and government processes of hundreds of organizations—organizations that by all rights should have crumbled under the tremendous inefficiency of their redundant, overspecialized, compartmentalized, costly business processes. There is only one reason why these organizations have survived—the same inefficiencies run rampant throughout their industries.

...competitive advantage, technology-based or otherwise, is temporary.

Perhaps I should not be so quick, however, to call the current practices of these organizations inefficient. After all, any competitive advantage, technology-based or otherwise, is temporary. In a free market economy, it's only a matter of time until competitive forces cause competitive advantage to become a standard practice for an entire industry. In this light we could say that efficiency, and the resulting productivity, is always a relative measure. In other words, inefficiency is problematic only when there is a benchmark for improvement.

This book is about a new benchmark; the level playing field has begun to shift—radically—and the benchmark for survival is changing. A few organizations are starting to pull ahead of the pack by applying the technology and discipline of automated workflow. These organizations are radically altering their work environments and, consequently, distinguishing themselves by achieving quantum competitive advantage in their process cycle times, product innovation, and customer responsiveness by double-digit factors.

This type of competitive disparity would normally dictate immediate change, but the *workflow imperative* represents a mandate for change greater than anything we have seen since the beginning of the industrial revolution, over two centuries ago. Change on this scale will rock the very foundation of current business practices and information systems, it will challenge common sense, and it will generate resistance from all of the interests so heavily invested in the status quo—no matter how inefficient the status quo may be.

Although the factory environment may seem a long way from the domain of the knowledge worker, there are some amazing parallels.

The cry for re-engineering ringing out through the hallowed hallways of corporations around the globe is an attempt to begin that change. Of itself, however, re-engineering is a noble but feeble effort that will only trade-in one problem for another. In the words of Tom Peters, it will "Replace old sterile organizations, with new sterile organizations."

The reality is that no re-engineering effort can cause an organization to change as rapidly as the forces of today's market demand. We have long since abandoned the industrial equation for productivity that inextricably linked growth to output volume. Bounding increases in factory efficiency allow today's global industrialized machine to produce more hard goods that can be consumed by the world's consumers. Productivity and growth in today's economic climate demands more than increased output. It requires the ability to innovate new products and services at rates that were unimaginable during the age of factory automation.

Although frustrating to organizations that resist change, the ability to respond quickly is becoming the key ingredient for success and survival. In short, success depends most on responsiveness. In the nanosecond nineties the prize goes to those who can respond quickly and, by the way, *constantly*.

Divorcing your organization from the "any color you want, as long as its black" legacy of the industrial age and making the transformation to mass customization and mass innovation means creating an adaptive organization that is able to constantly change—in intervals of days and hours, not months and years. Pilots refer to this as dynamic stability, the process of perpetually adjusting and compensating for environmental forces.[1] As in the video game Tetris, no single move in the change game constitutes success. Even the most obvious "right moves" create the platform for failure if they are not followed by constant and quick innovation.

So what does this have to do with workflow? The ability to quickly identify and respond to change in your market, economy, workers, and process is the fundamental benefit and the enduring advantage of workflow. Workflow is the culmination of the connected enterprise, constantly aware and constantly responsive to its internal and external environment.

---

[1] The inverse of Dynamic Stability is Dynamic Instability. In the case of the latter, slight movements are exaggerated in continuous increments as a small error is compounded into a major catastrophe. A possible equivalent scenario in business is the overcompensation in product line extension as an organization tries to reach an ever-diversifying market.

About this Book

The Workflow Imperative is about the inextricable relationship of business and technology. Each drives the other and both drive change. It does not dwell on which one is the initiator of change, and which one is the result; these are moot points to those of us tasked with managing the effects of change. (Few of us care much whether the chicken or egg came first—in either case we must all still get up each day to clean the chicken coop.) Instead, it will look at how technology and business disciplines have come together to form a new discipline—*workflow*.

Workflow bridges the enterprise, from manufacturing to the office, from technology to organizational culture. It is this unifying force that ultimately binds an organization, its people and processes together. In this sense, workflow has always existed in all organizations, whether it is automated or not, the flow of material, information, and knowledge must be orchestrated in order to deliver a product or service. Because there is no such thing as a single step process, workflow is always present, in some fashion to manage the pieces from step to step. But, this simple task, of managing the flow of work, is perhaps the single most important element of competitive advantage in mature markets, which have reached a stage of product, service, and positioning stability. At this point, competitive disparity can often only be diminished through quantum improvements in the redesign of underlying business processes. Add to this the global economic and competitive forces in today's business climate, and the automation of workflow becomes an imperative for survival.

To understand this phenomenon you must understand both the business impact and the technology of workflow. That is why this book is divided into two sections, one on each topic. *Workflow: The Business Imperative* looks at the ways in which workflow is already challenging and changing existing modes of work. *Workflow: The Technology Imperative* looks at the design and implementation methods, architecture, specific technology components of workflow, and the evaluation of workflow solutions.

If you already have a basic understanding of workflow you should read the two sections in that order. If, on the other hand, this is your introduction to workflow, you may want to skim through the technology section before coming back to the business issues discussed in the beginning of the book. In either case, reading both sections is the best way to appreciate the benefits of workflow's impact on your business, the correct design methods, and the fundamentals of workflow technology.

To understand [workflow] you must understand both the business impact and the technology of workflow. That is why this book is divided into two sections, one on each topic.

Workflow: The Business Imperative

An Opportunity for Change

The Lukewarmness [by which change is greeted] arises from the incredulity of mankind who do not truly believe in anything new until they have had actual experience with it.

Machiavelli, *The Prince*

The difference between visionaries and fools is just a matter of time.

T.K.

You don't need someone to tell you that your organization is changing. But we all need to be reminded just how much today's organization will have to change to cope in the coming years and decades.

Ever since the nineteenth century, the key indicator of success was productivity. We justified our investments and measured our success by the ability of workers to produce more with less—and for office workers the measure often consisted of just half of that—produce more, more, more. And now suddenly we are told, after two decades of rampant technology investment, that we have not increased our productivity at all! Who is lying here? Those who are working 60 to 80 hours a week, or the economists who have not yet learned that the fundamental formula for success has changed?

If the formula for productivity has changed, and growth alone can no longer be the measure of an organizations productivity, what then, is the measure? What is the justification upon which tomorrow's investments will be justified? Innovation—the ability to quickly, rapidly, and instantaneously respond to change. Many organizations already know this. For example, today 50% of Hewlett-Packard's sales come from products introduced in the last three years. Relentless, unyielding innovation—that is the nature of your business. And that will be the premise for every technology justification you make.

But it also means changing your organization at its very core to a new structure that is as dynamic as the global climate it inhabits.

In this climate of tremendous downsizing, organizations already are running lean. The problem for them is not to eliminate idle capacity and cut direct costs, but rather to cope with a flattened model after it has been imposed by existing business and economic conditions. Technologies will be justified based on whether they enable organizations to restructure on a continuing basis and empower their workers.

Perhaps a better way to look at the changes that have taken place in organizational structures is to define distinct periods of evolution for the modern organization. These boil down to at least four types of organizational structures:

The first and most familiar structure is the *Vertical Organization*. This relic of the industrial revolution is characterized by extensive hierarchies, approval committees, long decision time, resistance to change, and political strife caused by the isolation of hierarchical structures.

The contemporary "white collar" structure that many of us grew up with was that of the *Horizontal Organization*, characterized by matrix and team structures. Horizontal organizations are already flat. They rely on minimal management hierarchy and can respond quickly to certain decisions and investment, but, because they are flat, they are easily crippled by the lack of a communications infrastructure. They are also not immune from the politics of turf warfare.

The structure that has most recently been touted as that of the future is that of the *Virtual Organizations*.[1] A virtual organization is a modified form of the horizontal organization.

Although difficult to comprehend in the context of today's well-defined organizations, we can already see the beginnings of virtual organizations in those companies that adopt a recombinant structure in which resources can be quickly pulled together in a team to solve a particular internal or external problem. However, many of the participants in these teams are still functionally oriented.

---

[1] *Also see* 6 Epochs of Manufacturing, in *The Virtual Corporation*. Davidow and Malone, Harper Collins, 1992.

The benefit of virtuality is the reduction of organizational response time. The liability is the cultural impediment created whenever an organization adopts any structure—namely, that markets change faster than most organizational cultures are able to respond with similar change. Since this type of organization knows the value of responsiveness, justification is tied to time-based measures more than it is to direct costs. Justifying new technology is still problematic however, because the power base is still part of a static structure, albeit a very well-connected structure.

This is the fundamental reason why paradigmatic leaders, those who break the mold, have historically come from the ranks of small start-up companies and not from existing industry leaders.

The *Perpetual Organization* is the only constant structure that will survive change because it never stops changing. A perpetual organization can take the form of any structure, based upon the market demands at the moment.

How do we break free of the cultural impediments that intrinsically bound all organizational structures and create a perpetual organization? The key lies in understanding how an organization ultimately changes its form from vertical to horizontal to virtual. In almost every successful case it is the imposition of a subjective crisis, such as a CEO who anticipates a market trend, a large customer who demands change from a supplier, the expectation or realization of diminishing profits, or other quantifiable indicators of change. Notice, however, that I said "every successful case." The streets are littered with the remnants of organizations who realized their predicament long after any change could be instituted. Planning for change is much like blocking a penalty shot in soccer— if you wait until you know the direction of the ball, you have waited too long.

So how does an organization become perpetual, prepare itself for the unexpected, and react to events that have not yet taken place? Let's go back to the visionary CEO. He or she is given a certain amount of subjective discretion over the organization's direction and structure. Using this discretion, it may be possible for the CEO to justify redirecting the company at a crucial moment and, by doing so, realize a substantial success by entering the company in a new market. How often can one do this? Once a decade? Certainly. Once every five years? Perhaps. Once every year? Not likely. A CEO who restructures the company at every hint of trouble will be deemed to be either schizophrenic or just plain foolish, but hardly visionary.

There is another alternative. Constant and objective feedback from within the organization, accessible to decision makers such as the CEO and everyone else who participates in the process. This information would offer a constant feedback loop to help make decisions that create a perpetual re-engineering effort. This is not radical thinking. It is already being done in factories obsessed with tracking defects in the context of a TQM program. Retailers, such as Benetton, are using it to provide immediate feedback on product sales. Customer support organizations are using it to more evenly structure and utilize a workforce to handle customer calls that may change in makeup from day to day.

Management has always sought such objective information, but until now has been unable to get information about processes in real time. Workflow provides these metrics. With such metrics, initiating change becomes a matter of course, not an exception to the course. With such a culture established, an organization can open itself to a dynamic structure that embraces change through a constant stream of successful examples.

Admittedly, most people see incremental change as the ultimate sin. Why? Do we doubt the ability of functional workgroups and people on the front lines to develop an understanding of the importance of change and to use their best judgment to cause change?

Am I biased—you bet I am! Do I believe in empowerment? Absolutely. Does it work? No, not always, because it is not always preceded by education. If culture resists then it is for good reason. You do not negotiate change. The only leverage you have to implement change is proof that it will succeed. The only proof you have is a track record, and the only way to establish a track record is to start somewhere—anywhere—if it is a decision between starting and going nowhere.

Even when all is said and done (admittedly, *done* is the operative word) you will still have to face a variety of obstacles to flattening an organization and creating a perpetual model of change.

As you embark on your change efforts remember that the tools you use to justify and enact change are just that—tools. The way they are used may change from one organization to another. Methodologies are not meant to be cults; they are simply guidelines. Be leery of anyone who tells you otherwise.

An education awaits you, and the only way to begin is to start today. Start large, start small, but start, and don't expect to ever end.

Workflow: Agent of Change

Workflow is more than just an approach to managing change. It is also a specific set of methods and technologies that represent a massive swing in the tools and methods used to support a business process. So massive, in fact, that it is doubtful any organization will be able to keep pace with the metronome of change if its processes are not enabled with these new technologies and methodologies. Not unlike the tremendous competitive pressure to invest in factory automation and quality assurance techniques during this century, workflow systems will become a cornerstone of competitive advantage.

Such an edge is desperately needed in the white-collar work force, which, despite the monumental investment in information systems, has apparently yielded no increase in productivity over the past two decades. To the typical office worker this fact seems to fly in the face of constantly increasing office output. Consider how much more work you can do today in any given period of time as compared to the same period of time five years ago; but the increase has not been without significant investment in office technology. The result has been a net productivity gain in the white-collar work force of less than one percent over twenty years. It is clear, as we already said, that output is no longer the measure of successful organizations.

Although the factory environment may seem a long way from the domain of the knowledge worker, both are part of the same value chain of activities. Within the factory we have already embraced a new information-based paradigm, where techniques such as mass customization and just-in-time production are achieving economic order quantities of single units. The mass production legacy of the last three centuries, however, is a constant progression toward ever-increasing work fragmentation and specialization, this meant that in the office, each new pairing of specialized process steps required a corresponding element of time and effort to transfer information from one step to the next. Miscommunication thrived, as specialization lead to more people and more points of control in the process.

The modern office is still very much like a Minoan palace that winds and turns in myriad directions. Specialization and highly distributed operations have created monstrously complex interactions between knowledge workers, and that complexity coupled with the advent of cost-effective connected desktop computing has created an enormous competitive opportunity to collapse and simplify these complex, protracted business cycles.

The walls we have placed between the parts of an organization, its people, its suppliers, customers, and all participants of the value chain are artificial. Bringing productivity gains from the factory to the front office should not be regarded as a leap across a divide, but rather a bridge that acknowledges the importance of bringing discrete processes within organizations closer together. The ability of workflow to bring these pieces together will change the fundamentals of the office. Not only will technology and organizational infrastructure change, but the very nature of work will be transformed.

It is certain that transformation of this scale will not come without a human cost and an added measure of social responsibility. We have seen vivid evidence in the drastic impact of downsizing on our blue-collar workforce, and now we are beginning to see the same effects in the front office. When technologies are used only as a means to accelerate downsizing and increase productivity, they do not also increase prosperity. If we ignore the call to social responsibility, and do not apply process automation technology to enhance the role of the office worker,[2] we will see social upheaval that pales that of the industrial revolution. This upheaval may be inevitable regardless of our actions. Downsizing of the white-collar workforce continues at increasing rates, with which our social institutions are ill equipped to cope. We are entering what appears to be the breakdown of dynamic stability, a spin. Inexperienced pilots can rarely recover from this situation—and it remains to be seen what we will do to re-integrate millions of newly unemployed white-collar workers into our workforce.

Idle promises and threats? It may be too early to tell with certainty, but the fate of the office worker and the transformation of the office environment through methods and technologies such as workflow will be more dramatic and complex than we can anticipate, such is the definition of any paradigm shift. After all, anyone who predicted fifty years ago the enormous changes brought about in the factories and production lines of Japan certainly would have been considered crazy. Today these people are called visionaries.

"Workflow" is a tool set for the proactive analysis, compression, and automation of information-based tasks and activities.

Defining Workflow

Workflow applies many of the same concepts and benefits of factory automation and industrial engineering to the process of work management in an office environment. We define workflow as *a tool set for the proactive analysis, compression, and automation of information-based tasks and activities.* The basic premise

---

[2] Although we will make reference to office workers and office automation, it should be made clear that workflow automation applies not only to office settings, but in many cases is the technology that bridges the two communities of office and factory workers, shortening the business cycles between them. In fact, faster market response, for example, developing and delivering a new product, hinges on the ability of marketing and manufacturing to work in close harmony, not as separate and isolated business units.

Although the factory environment may seem a long way from the domain of the knowledge worker, there are some amazing parallels.

of workflow is that an office environment is an information factory, or more specifically, a process factory. The process, which can exist in a range of formats from paper to electronic form, provides the basic raw material of every office task. The connection of these office tasks creates a *value chain* that spans internal and external task boundaries. In this architecture, workflow attempts to streamline the components of the document factory by eliminating unnecessary tasks, thereby saving time, effort, and costs associated with the performance of those tasks and automating the remaining tasks that are necessary to a process. (The value chain concept is explained later in this chapter.)

Treating computerized office systems as a commodity has made this type of analysis and change possible due to their evolution from isolated personal productivity tools into networked resources for electronic process transfer across organizations. This trend towards connectivity among office workers will undoubtedly prove to be one of the most significant events in the history of white-collar productivity.

There are those who would leave it at that and say that workflow is simply the natural evolution of interconnected desktop computing, but much more is involved in the popularity of workflow.

The traditional organizational hierarchy, with its vertical emphasis on communications, has resulted in information systems that do not support the horizontal nature of collaborative, or team-based, communications. In reality, we all know that processes do not much care for the arbitrary boundaries of a compartmentalized organization. Instead they must traverse an organization's infrastructure both vertically and horizontally. In addition, as organizations embark on re-engineering efforts that eliminate over-specialization in the workforce, the underlayment of the industrial revolution, a new breed of generalist is evolving. The new generalist is one who is no longer constrained by hard and fast departmental boundaries. These generalists work together in extended coalitions of workers that cut across an organization's structure, geography, and politics. No longer process components, these individuals are now process owners, and in increasing numbers are becoming owners of the business, as well.[3]

These extended coalitions of office workers have created a new work environment of enormous complexity and interaction. CEOs share electronic dialogues with salespeople, customers become part of the process flow, and processes are amalgamated to create hybrid systems, along the lines of mass customization, intended for the delivery of a specific and optimal solution to every problem. Such complexity requires new tools for the coordination of activities and communications. The foundation of these tools is workflow technology, which extends beyond the realm of vertical information management to that of horizontal process management.

Redefining the Information Asset

One of the most important aspects of workflow's advent is the impact it has had on the concept of information as an organizational asset. The premise that information is a tangible asset that must be

---

[3] One of the most pronounced trends in employee empowerment has come in a tangible form, that of equity participation. In many instances this type of ownership is an attempt to align employee success directly to that of the company; as may be the case for many small businesses. But an increasing number of large organizations, such as Avis and United Airlines, are being converted, through such programs into employee owner organizations. Although ownership, at these levels, may not appear to have an overwhelming influence, it is an outward, tangible manifestation of the increasing emphasis being placed on ownership over process.

preserved and valued does not change. What does change is the concept of *information* management. In the traditional view, information management was represented by the data and documents that were used in the support of a business process.

With workflow we add a new dimension to information management—that of the process asset. This is also the principle difference between workgroup computing and workflow. Workflow enables organizations to capture not only the information but also the process, including the rules that govern its execution. These rules include schedules, priorities, routing paths, authorizations, security, and the roles of each individual involved in the process.

Before chords of fear begin to resonate for those readers who see this as yet one more way to downsize the workforce, they should stop and think about the flip side of downsizing. All too often we attribute downsizing to technology. Though technology may enable downsizing, it is typically external economic and global factors that drive the downsizing, cost-cutting, and re-engineering of organizations. Whether we like it or not, most organizations will have fewer office workers burdened with performing more and more tasks in less and less time.

Without a means by which to validate, compress, and automate work roles, downsizing has no long-term benefit. Downsizing alone would send industry into a death spiral of ever-increasing inefficiency as workers are burdened beyond their abilities and capacity to cope. Workflow is a survival response mechanism. It not only stems this process but can actually improve the standard of work environments by minimizing miscommunications among heavily burdened work groups.

It is also worth pointing out that if external factors were different, the perception of workflow as the knot in the hangman's noose, would be equally different.

With workflow, we add a new dimension to information management— that of the process asset.

For example, during a boom period of abundant opportunity—think of the 1980s—any technology or method that helped expand capacity and convert opportunity to revenue was extolled as beneficial, even though it often raised headcount and costs—a legacy we are now working hard to expunge.

Instead, the technology infrastructure for workflow did not come into existence until the economic climate was substantially changed from the growth-or-bust attitude of the 1980s. The phenomenon of automation as a tool to restructure organizational work environments, so closely parallels the automation and productivity increases in manufacturing during the second half of this century that we will use the term "process or document factory" in this text to refer to office environments. A caution is in order, however. The factory provides a strong analogy that helps us to better understand and discuss the problems and benefits of workflow. But we will see that it is also dangerous when this metaphor is adopted in its extreme form.[4]

Whether we like it or not, most organizations will have fewer office workers tasked with the ever-increasing burden of performing more and more tasks.

Workflow Versus Workgroup Computing

One of the most often asked questions is, "What is the difference between workflow and workgroup computing?" Given the tremendous market opportunity that each provides to technology vendors, it is hardly surprising that the differentiation is blurred by the plethora of offerings that are using both labels. The distinction is actually a simple one. Workgroup products facilitate the transfer and sharing of information from workgroup to workgroup or individual to individual. The key ingredient is the *information*. In a workflow application the process

---

[4] The factory analogy is important from the standpoint of its context but it does not hold true in practice as we will see later.

Workflow versus Workgroup Computing

Workgroup computing (top of page) focuses on the information being processed, enhancing the user's ability to share information within workgroups. Workflow (bottom of page) emphasizes the importance of the process, which acts as a container for the information. In this way workflow combines rules, which govern the tasks performed, and coordinates the transfer of the information required to support these tasks. This is a "process-centered" model, as opposed to an "information-centered" model.

Workgroup Model: *Information Centric*

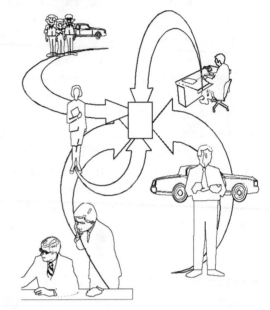

Workflow Model: *Process Centric*

knowledge that applies to the information is also managed, transferred, shared, and routed. The key ingredient is the *process*.

This may appear to be a subtle distinction. It is not. Process knowledge involves capturing the roles, schedules, and resource descriptions and then automating these as part of the workflow application. Many workgroup applications provide a minimal level of process functionality, for point-to-point workflow or ad hoc processes, similar to an E-mail system, but only workflow products focus on the issues and problems inherent in process automation, such as the analysis of processes and the definition of role relationships. (For example, how does a credit analyst work with a sales person, in any one of a dozen interactions?)

The Document Factory

An office environment thrives on a measure of creative chaos.

A revolution occurred in factories as assembly lines consisting of discrete tasks gave way to tightly coordinated parallel processes of manufacturing, inventory management, and retooling for agile manufacturing and mass customization. Workflow will similarly revolutionize the document factory as it creates an environment for the streamlining and integration of information-based activities at never-before-possible levels of integrity and reliability. These last two points are significant in an era of downsizing organizations with smaller workforces and greater competitive pressure, as measured in the form of shorter time-to-market and increased customer responsiveness.

Any discussion of an office factory will, however, be met with skepticism and outright hostility on the part of many. As was mentioned earlier, the factory analogy has its shortcomings when applied in a literal sense to an office environment, as do the methods and tools of industrial engineering.

Factory environments can be highly structured and disciplined. Tasks occur with relentless precision and consistency. Errors are reduced to zero defects and hair-width tolerances. In a factory, this type of environment can be maintained for long periods of time without flaw. In an office environment this same approach will soon result in reduced productivity and lost opportunity as knowledge workers lose the ability to exercise judgment in structuring their own time and creative processes. An office environment thrives on a measure of creative chaos. Periods of high productivity are often preceded and followed by what would be considered idle time on the assembly line. If managers and developers of workflow systems do not recognize these fundamental differences, they will not only frustrate users, but also undermine the potential for increased productivity.

We can acknowledge the merits of this argument and accept that there are some irreconcilable differences in the nature of work from factory to office. At the same time we should also accept that there are similarities, and if we are to streamline any organization we must somehow bridge the distance that for too long has stood between these two communities.

Once again, however, it is not technology that is mandating this profound change in organizations, but rather external forces. One of the most pronounced of these is the relationship between productivity growth and the health of a free-market economy. Although global economies have experienced many fluctuations during the twentieth century, there has been an average 1.5% annual increase in the productivity of the world's industrialized nations since the turn of the century. Almost all of this has come from improvements in the unskilled labor force, namely farm and factory workers. We have been so successful at engineering factories that, according to many, our ability to provide goods has outstripped demand. As stated earlier, the world's manufacturers now provide 130% of all the goods required by the markets of the world. We find ourselves at an anomalous and unique point in history when, even if manufacturing productivity continued to increase as measured in units of output over units of input, it would not necessarily increase prosperity (if we define *prosperity* as the ability to derive greater value for increased productivity).[5]

This is not an unsolvable predicament. The solution is simply to produce goods faster and smarter, not simply to produce more goods. The faster and smarter aspect of the equation is primarily the responsibility of the office staff. These individuals are the innovators. The faster they can innovate, the faster the right products can be brought to the market. Increasing productivity no longer means just increasing output; it means increasing value by accelerating the collaborative and creative processes that lead to innovation.

At this point those readers familiar with traditional tools and disciplines, such as project management and time-motion studies, may ask if these are not adequate for use in automating office environments.[6] If these same individuals, who are well-versed in the methods and purpose of such tools, were the only ones involved in defining business processes, the answer would be yes. The reality, however, is much more complex.

---

[5] Productivity is often calculated as output divided by input. A more precise measure is the change in value provided over a series of time periods given input and output costs, and value provided for each period. So, for example, if one fully burdened factory worker costs $70,000/year and produces $140,000 in goods in year A and then, at the same cost of $70,000, produces $210,000 in year B, we could say that the factory workers productivity increased by 100% from year A to year B. ($140,000 [year one output] - $70,000 [year one input] = $70,000 [year one value produced] : $210,000 [year two output] - $70,000 [year one value produced] = $70,000 [change in value produced or productivity change/increase] : $70,000 [year one value] / $70,000 [year two increase] = 100% [productivity increase]. If, however, we were only able to sell 66% of the goods produced in year two there would be no increase in productivity! $210,000 [value of goods produced in year two] * .66 [actual goods sold in year two] = $140,000 [year two actual value of goods produced] - $140,000 [year one actual value of goods produced] = 0 [or, no change in productivity even though there is a change in output.

---

[6] This reference to "tools" focuses on the appropriateness of the analytical tools but does not yet bring their validity into the discussion. We will argue later that many of the mechanics used in a production environment, such as the measurement tools proposed by Fredric Taylor are inept and arbitrary in a knowledge-based work environment.

The tools and methods required for business process automation must involve not only analysts and developers, but more importantly, the users and sponsors of the business systems. These people are not skilled in industrial engineering, project management, or systems analysis, and they have little interest or time to develop these skills.

The answer is the development of high-bred tools that are accessible to developers, users, and sponsors alike. Unlike traditional tools and methods that were entrusted to the domain of professional analysts, designers, and developers, these tools are part of the workflow solution. This creates an environment of collaboration and responsiveness between developer and end user, which was not possible with the use of traditional tools. The effect of this synergy will be especially apparent when you read the case studies in this text.[7]

It is also important to realize that this sort of collaboration requires ongoing commitment on the part of sponsors and end users. This means taking care not to use the workflow's capabilities for providing process metrics as a tool for individual persecution. Whether the metrics are accurate or not, office workers are particularly sensitive to this type of scrutiny, as well they should be. The productivity and the value of a knowledge worker is much harder to measure than that of the factory worker, who typically produces a finite quantity of a tangible product. The unit-of-work in an office is not always identifiable, and when it is, a good measure of judgment and creativity is involved in the creation of its value. This ambiguity makes the

---

[7] The Canadian Broadcasting Company (CBC) case study, later in this text, attributes much of the success of its re-engineering to the level of communication between planners, implementors and the people who actually use the system. This type of collaborative design and development requires tools that permit a high level of interaction during all phases of system concept through system operation.

true value of workflow much harder to define for many organizations who are using work flow as a tool for the measurement of people efficiency.

One of the most astounding insights for those new to business process redesign is that even the elimination of all "people" task time will only reduce overall process time by a small percentage. The task time of a process typically accounts for only 10% of overall process time, the other 90% is transfer time. This is one of the principal reasons that workflow should initially be used as a tool for measuring "process" transfer time. In this way, the focus is shifted from the people to the process, which is not a subtle shift.

At some point after transfer times have been reduced or eliminated through a sound analysis, it is not unthinkable that you may find yourself reconsidering the tasks themselves. Starting an analysis by pointing a finger of blame at the same individuals you want committed to the solution will, at the least, cast doubt upon your intentions.

Finally, before you begin to evaluate the potential of workflow and draw analogies to a factory-based model, you should learn to distinguish knowledge work from production work. Both knowledge workers and production workers are essential to any organization. Although both are valuable, the level at which workflow has immediate value to an organization tends to be closer to the production end of this spectrum.

Production work, such as a mailroom activity, can be broken down into finite, discrete, and easily defined tasks. These tasks are both repetitive and predictable. True knowledge work involves spontaneity and judgment-based reasoning that evades the ability of any automated system to capture and preserve the rules governing a process. Automating this type of workflow will often take longer than the tasks themselves. Since each task may be a one-time exercise, there is little if any apparent value to their automation. See the section on *Time-based Analysis* later in this book for details.

This is not to say, however, that you should exclude knowledge workers from a workflow process. Instead, you should involve them fully in the definition and use of a workflow solution, but their components of the process will almost always require the less-structured aspect of a groupware solution. Well-defined workflow can be combined with an open ended groupware function, whether through E-mail or other means. This will best meet the needs of processes that rely heavily on the involvement of knowledge workers.

The factory productivity phenomenon can be replicated in the office...

Despite all of these fundamental differences between office and factory, factory productivity can be replicated in the office by applying some of the same principles that were used on the factory floor, but with a slight modification. Factory automation disciplines such as Total Quality Management (TQM), Electronic Data Interchange (EDI), just-in-time (JIT) methods, and Time-based Analysis are inherent in workflow technology. With a few cornerstone tenets in place—the necessity of creative chaos in an office environment, the valid sensitivity of office workers to performance metrics, and the distinction between knowledge and production work—we can begin to look at some of the process components that are shared by all aspects of the value chain, from the front-office to the assembly line. We will explore these areas in the following pages and, as we do, break down the barriers from factory to office—barriers that have stood in the way of bringing the steady increase in manufacturing productivity across the great divide into the office.

TQM and Workflow

TQM is a broad-based model for evaluating and redeploying every aspect of the organization towards the final goal of increased service and product quality. Information moves quickly and directly to the individuals closest to the process. The presumption is that these individuals are best able to determine what problems may exist and what measures need to be taken to improve quality in a timely fashion.

Workflow facilitates TQM by providing tools to bridge any gaps between layers of management.

Workflow facilitates TQM by providing tools to bridge any gaps between layers of management. In fact, in some cases the approach is so effective that the ultimate result is a flattening or restructuring of the organization to eliminate layers of inefficiency. In these new organizational models, workers are empowered with information and the authority to actively contribute to the quality process.

In turn, the TQM environment generates a much higher need for ongoing communication and coordination at a peer to-peer level because the traditional management hierarchy no longer provides top-down coordination. Tools such as workflow ensure the integration of these tasks through automation of the information transfer process. The benefit is improved communication and better responsiveness. The result is a more competitive model for a quickly changing marketplace than that provided in an extensive and tedious management hierarchy.

Workflow in the Extended Enterprise

Recent advances in technologies such as EDI (Electronic Data Interchange) have brought about a new model for describing the structure of an organization's information systems. This new model, called the *extended enterprise*, has far-reaching implications on

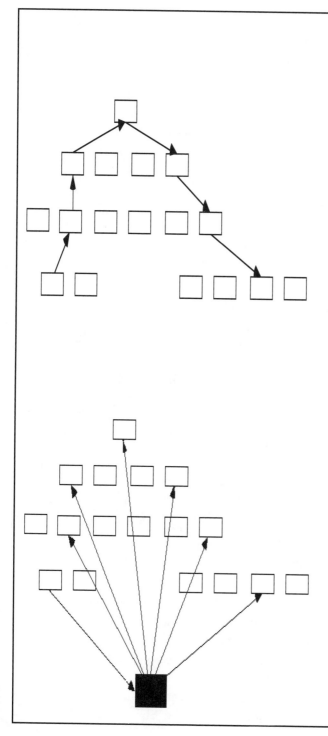

TQM Bridges Management Layers

The traditional path of information (top of page) that is used to initiate change in an organization follows strict organizational boundaries. This can lead to delayed response to problems, reduced quality and service, and extended business cycles. Worst of all, any response taken may be too late to impact the original problem, which initiated the change request.

In the TQM/ Workflow model, information is delivered directly to those who need to take action, while it is also broadcast to those who need to be aware of the change and action taken. This model empowers those closest to the process without eliminating the organizational structure.

workflow technology because it causes the workflow application to extend beyond internal data. In the extended enterprise, an organization's information systems are no longer bound to internally generated data and documents. Instead they are redefined to include all documents and information that apply to a particular task or process. This could extend well beyond the enterprise into outside organizations, such as suppliers and customers.

The Extended Enterprise Model at Work

Extended enterprises operate as though they where one enterprise. In a manufacturing process, for example, Company A's manufacturing process requires parts from B. B's inventory is accessed by A's workflow. Since B has low inventory, B's workflow will trigger a request for raw materials to C. C fulfills the request for B and A receives its parts.

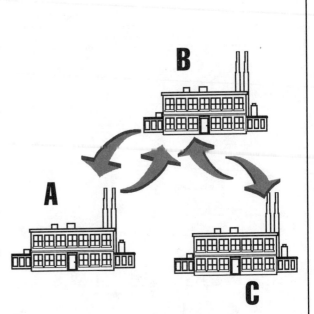

A prime example of the extended enterprise is the use of EDI in situations where business information is shared among trading partners, across organizational borders, to expedite transactions. Trading partners share certain common information with access privileges that allow direct access to each other's information. Through the EDI extended enterprise model, an order placed by Ford for tires supplied by Goodyear would give Ford direct access to Goodyear's inventory information. Ford may be able to initiate a manufacturing process at Goodyear that corresponds to Ford's manufacturing schedule based on production times and schedules at the Goodyear factory. This may trickle down to Goodyear's raw material supplier who is indirectly defining Ford's manufacturing schedule based on seasonal inventory cycles.

The challenge posed by this model is the establishment of parameters and rules for the integration of the two business partners manufacturing activities. Workflow becomes a major factor in establishing an effective extended enterprise model. Workflow provides the tools for defining these parameters, the integration of discrete applications, and the rules of any extended enterprise transaction. In fact, the workflow model would work just as well with a system that integrates office functions, such as sales and marketing, as it does on the shop floor.

Value Chain Analysis

A valuable framework for considering how workflow systems can contribute to management and organizational effectiveness, in either an extended or a traditional enterprise, is based on the concept of the *value chain*. The value chain provides a specific framework for the assessment of the critical areas that can be facilitated by the automation of workflow. Originally developed by Michael Porter, professor of management at Harvard Business School and author of *Competitive Advantage*, the value chain concept has become widely used in studying the strategic impact of new technology on the business model.

Although the original work done by Porter was not developed specifically for workflow technology, the principles and the approach of value chain analysis can help determine those areas where a workflow system facilitates the critical transformation of data into information and knowledge, which in turn can yield major strategic benefits. We have taken the concept of the value chain and applied it, with some modification, to the task of assessing a series of activities that deal with workflow.

The value chain views any organization as a collection of activities that add value to the product or service of an organization. For example, value is added when the production process takes raw materials and transforms them into a finished product, when the product is distributed, and when the product is sold to the consumer. A company is profitable if the price consumers are willing to pay for a product exceeds the costs of creating value.

The value chain is a system of interdependent activities connected by linkages. Linkages occur when the performance and execution of one activity affects other activities. For example, the ability to share clinical data among researchers within a pharmaceutical company may be necessary to develop an application for review by the FDA. The cost of creating value is a function of how well the activities in the value chain are coordinated and integrated. A poorly structured value chain may result in delays during the transfer of information from one task to another, causing the time to market for a product to expand beyond a window of opportunity. This could be a significant loss of opportunity for a pharmaceutical company bringing a new drug to market.

External Linkages

Besides these internal linkages, there are also external linkages to other organizations such as suppliers or distributors, as we saw in the discussion about the extended enterprise. Just-in-time (JIT) inventory systems that involve significant information sharing between a company and its suppliers are a prime example of how a workflow system can enhance such linkages.

By networking an organization with its suppliers and distributors, improved delivery, inventory reduction, and other key benefits can be realized. In this value chain, the workflow system acts as a means by which the optimal coordination of tasks and activities between the two trading partners is ensured.

Applying the Value Chain Model

Gaining sustainable competitive advantage is the ultimate purpose of introducing workflow into value chain activities. In assessing the strategic significance of workflow applications, companies should consider the following:

- Will the workflow system so transform a value chain activity such that it will create barriers to entry for competition (i.e., through lower costs or increased differentiation)?
- Can the workflow system intensify and enhance customer relations through more efficient sharing of information?
- Can the workflow system change the basis of competition (by shortening the business cycle or reducing production and/or service costs)?
- Does the workflow system allow for the creation of strategic linkages to business partners and suppliers?

These questions should be applied to the value chain analysis of the company, its customers, known competitors, and potential new entrants. For example, the use of workflow systems in the research and

*Identifying your
value chain of
activities need
not be a complex
process.*

development (R&D) value chain activity might create barriers to competition by making it difficult for new entrants to replicate extensive and sophisticated workflow models that result in shorter development times. If such a system also incorporates linkages to other research systems, this competitive advantage will be notably strengthened. Careful coordination of linkages can be a powerful source of competitive advantage because potential rivals have a more difficult time detecting and understanding such linkages.

Identifying your value chain of activities need not be a complex process. Although the value chain of an entire enterprise may be very intricate, that of a single workflow process can be easily defined.

The first step in the process is identifying the overall information infrastructure of your organization. This should include a definition of all existing computer-based and manual information transfer operations. Take nothing for granted in defining all the linkages of your organization. It may be that there are informal means of information transfer that are critical elements of the overall system, and these may be some of the least efficient aspects of your current workflow.

It is best to do this with a graphic representation of the information system. There are formal methods for accomplishing this, such as the System Schematic methodology. (The System Schematic is described in more detail later in this text.)

Once you have a working diagram of the present system, the second step is to identify the specific contribution level of each component in the diagram. These represent the individual links of the value chain. This is done simply by asking, "What has this component of the overall workflow contributed to the organization's profitability?" In other words, what is its value? Recall here the discussion earlier about the importance of assessing value based on more than just output, which is why our focus is on profitability. The correct approach uses a zero-based method of justifying the value of the existing information systems and activities. The premise is that information systems are not a "cost of doing business;" instead they should be an active contributor to the value chain of an organization.

This is where a thorough value chain analysis becomes an especially useful tool for assessing the impact of workflow. It is likely that you will find activities that do not actively enhance the organization's value chain. All too often it is taken for granted that these activities are a necessary cost of doing business. By assigning a value to each activity you can identify the quantifiable benefits of implementing a new approach to managing a specific workflow.

The premise is that information systems are not a "cost of doing business;" they are an active contributor to the value chain of an organiztion.

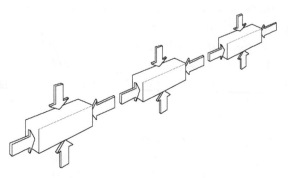

One of the most pronounced benefits of workflow is the elimination of idle transfer times.

The Impact of Time-based Analysis on the Value Chain

The most important element of value chain analysis, and one that is most often ignored, is *transfer time*. (We will also refer to *queue time*, which is one element of overall transfer time. The other is *physical transfer time*.) In every office environment two elements affect the overall efficiency of a business cycle: *task time* and *transfer time*. Although either one can slow down a process, transfer time is usually the worst culprit and best-hidden villain in an organizational setting, with 90% of a typical business cycle being attributed to transfer time.

Transfer time is customarily ignored during an analysis of existing procedures because the focus tends to be placed on the people and the tasks, not on the time that passes between the completion of individual tasks. Workflow, on the other hand, addresses both task and transfer time and can effectively collapse the idle period between tasks. Although the result may appear to be just a function of using electronic information as opposed to paper-based communications, it is far more involved.

Electronic information can be accompanied by many of the same problems that plague paper-based systems. For example, a large East Coast insurance company was surprised to find its pilot electronic imaging system did not expedite claims processing. On closer inspection, it turned out the claims processors had not adapted to the new workflow model. Rather than work on claims individually as they were received, the claims processors would wait until a week's worth of claims had accumulated before processing a single one, just as they had done with the manual system. Had the company performed a Time-based Analysis, they would have realized that their existing processes were inefficient, due not so much to the issue of task time, but to the transfer times inherent in the existing business process. Processes could have been redesigned taking full advantage of the capabilities of workflow and imaging technology by understanding these differences between task time and transfer time.

It is also important to appreciate the difference between physical transfer time and queue time. In an electronic system only the physical transfer time is eliminated. Queue time, the time an activity is idle while waiting for work to be performed, may be the dominant component of transfer time. If that sounds hard to accept, consider the amount of time that work spends in your IN or OUT basket. Unless they are always empty, you are part of a process that includes both physical transfer time and queue time. In fact, one of the techniques you can use when analyzing transfer time, in a paper-based system, is preprinted pads of yellow sticky notes that are attached to documents to track transfer time.

If transfer time is not clearly understood, tasks may be substantially reconstructed, at significant expense, with no measurable increase in the efficiency of the overall business cycle. Consider, for example, that a business cycle, that has a ninety-to-ten transfer-time-to-task-time ratio would be shortened by only 10% if *all* task time were eliminated. This is an obvious fact, yet one that is often missed by those who focus exclusively on modifying tasks, the behavior of workers, and individual productivity.

We will look at the issues of task and transfer time in more detail in the workflow methods section of this book. For now it is important to understand the basic difference between the two, and the dramatic impact that transfer time has on the efficiency of a business process.

In most cases the more removed from the next step in a process, the more time we spend trying to support the process...

The most critical liability created by extensive transfer times is in the loss of intimacy between participants in a process. *Intimacy* may sound like an odd term to use in this context, but think about its meaning. To be intimate with someone or something connotes familiarity beyond the obvious. It also means that information can be conveyed, albeit a one-way conveyance in the case of an object, in an informal manner.

Now think about the way in which you exchange information with various colleagues in a business setting. Is it not true that those individuals whose processes are best known by you are also those with whom you have the highest level of interaction? Naturally, this leads to an understanding of what they do and why they do it that exceeds their job descriptions or the text in a policies and procedures manual.

Now consider the time that elapses between interactions with colleagues with whom you are less familiar. With which group are your interactions better documented, for example, by memoranda? With which group do you typically exchange more information about a process and its rationale? With which one do you exchange more information about the *information*? In most cases the more removed from the next step in a process, the more time we spend trying to support the process, the less our understanding of the process, the greater our emphasis on the obvious element of the information object, and the less likely our exchange of informal knowledge about the process. All of this results in less intimacy.

So how do we increase intimacy and bring people closer to the process? By deflating the transfer time involved in a process. Transfer time undermines process understanding by isolating tasks. The intimacy fostered by informal communication and observation, which is such an important part of communicating beyond the self-evident, is lost. With it we lose the spontaneity and informal paths of communication which are invaluable to building a responsive organization.

I am not suggesting a return to centralized organizations; rather, use of workflow offers the ability to collapse transfer times and increase process intimacy by communicating through a separate path that does not rely on the direct communication of one task to another.

Consider the analogy of an inflating balloon. Envision tasks as dots on the surface of the balloon and transfer time as the lines that connect these dots. An inflating balloon can be likened to a process of increasing complexity, specialization, and growth. As the balloon inflates, the transfer times increase and the tasks become increasingly isolated.

The only solution is to reducing cycle times is to communicate along lines that are not only on the surface of the balloon, but instead traverse the process outside of these two dimensions. Bear with me, I am not going to suggest we traverse the space time continuum through a fourth dimension, but there are creative alternatives, such as the asynchronous communications model discussed later in this text, which can substantially reduce cycle times by providing for person to process communication.

The bottom line is that any method which decreases transfer time will increase intimacy. As for defining the "right" amount of transfer time, there is no single measure. Even though a one-to-one ratio may be exceptional for business processes that previously required at ten-to-one transfer-to-task ratio, this is only a relative measure. Recall that it is the definition of the current benchmark and the value added by changing the benchmark that need to be considered, not the absolute value. Also remember that communication needs to be defined as time to a result, and this rarely happens on the first iteration of communication. That means that even the single round trip transfer time is a worthwhile benchmark to shoot for, especially in process cycles that require numerous negotiations.

Few of us would argue that intimacy is needed for strong and secure interpersonal relationships. Why then should we doubt that they are necessary for the communications that take place between the individuals who comprise a business process? Ultimately, achieving this level of communication between individual participants in any business process determines the success of an organization.

Human Factors and Re-engineering

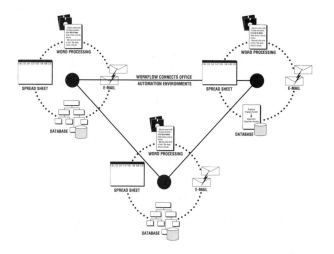

Workflow often represents the critical link between technology and people.

It can be said rightfully that workflow brings little to an organization that is not already there in existing office automation technologies. What workflow does is consolidate existing office automation and document management technologies under a single environment. It is, effectively, a complex distributed data manager.

Workflow often represents the critical link between technology and people. When it is missing, the whole system suffers. The conversion to electronic documents does not add in and of itself the intelligence and control necessary to achieve the benefits of process automation, nor does it necessarily expedite a process. Workflow orchestrates the fundamental components of a business process: roles, rules, and routes (think people, applications, communications). It is this uniting quality of workflow, the workflow environment, that is most often misunderstood.

The idea behind workflow technology is to create a single environment to manage the complexity of multiple office automation environments. As software and data have migrated from individualized solutions with dedicated functionality to integrated solutions then to groupware solutions, workflow has evolved as a metaphor for the coordination of multiple workgroups using multiple technologies. In many ways, workflow becomes the conductor of data, documents, applications, communications, and the user interface.

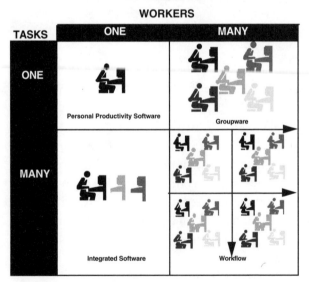

Initially, however, workflow enters an organization as a simple two-dimensional solution, handling multiple technologies used by multiple individuals within one or more workgroups. In time, the system is expanded for use by many different workgroups, transforming it into a three-dimensional model. For many organizations, this may be one of very few opportunities to share electronic information and perform electronic transactions across workgroup boundaries in a single cohesive environment. The complexity of the workflow model grows quickly at this stage, as user requirements and the diversity of information sources and applications increases. This three dimensional metaphor is important to understand in planning for workflow. Organizations must carefully select workflow products with an adequate range of tools to support a variety of office automation technologies and allow for the development and orchestration of this diverse environment.

Before considering the specific technology requirements, however, it is important to understand the workflow environment from the standpoint of its impact on business processes and the human dynamics of the organization. The analysis of these issues must precede any technology evaluation or discussion, since they apply not only to workflow technology but also to the broader realm of process redesign. We will look first at the cultural issues and second at the relationship of workflow to re-engineering.

The largest single obstacle faced by organizations planning to implement workflow is that of existing organizational culture.

Human Factors, Corporate Culture, and Workflow

Unfortunately most information professionals do not easily accept the fact that the real issues standing in the way of workflow's adoption are not technology-based but are mired in human factors and organizational issues. The reality is that study after study shows that in more than 50% of all cases culture is the largest obstacle identified by

evaluators and users of workflow.[1] If these issues
are not addressed, the success of workflow is simply
impossible.

The cultural effects workflow can have on the
organization are numerous. They include: the
flattening of the organization's management
structure, the imposition of a new level of control
over office workers, increased availability of
information to the point of overload, and diminished
proprietary interest in processes. If not given careful
consideration at the outset, these issues can be
devastating to the long-term viability of the
workflow system.

Workflow is a tool for adapting to the already changing structure of the organization.

Flatter Organizations

Flattening of the organization's management
structure is a mixed blessing, as we have already
seen in our discussion about TQM. It is not however,
an inevitable result of implementing workflow
technology. In the climate of downsizing that is
prevalent today, organizations are already running
lean. The problem for them is not in eliminating
idle capacity and flattening the organization, but
rather in coping with this model after it has been
imposed by existing business and economic
conditions. In this regard, workflow is a tool for
adapting to the already changing structure of the
organization. It enables organizations to restructure
more effectively and preserve stable market share
and business volume with reduced staffing.

Perhaps a better way to look at the changes that
have taken place in organizational structures is to

---

[1] In 1993 Delphi conducted a study of 400 organizations
evaluating or implementing workflow solutions. Over 60%
stated that the primary obstacle to workflow's adoption was
cultural and not the immaturity of technology, steep costs, or
lack of standards.

define distinct periods of evolution for the modern organization. These boil down to at least four types of organizational structures:

- **The Vertical Organization**— Characterized by extensive hierarchies, approval committees, long response times required to send information up and down through the hierarchy in order to make even basic decisions, resistance to change due to the inherent justification process necessary for investment, and political strife caused by the isolation of hierarchical structures.
- **The Horizontal Organization**— Characterized by matrix and team structures. Horizontal organizations are already flat. They rely on minimal management hierarchy and can respond quickly to certain decisions and investments, but they are easily crippled by the lack of a communications infrastructure. They are also not immune from the politics of turf warfare because the teams often form long-term functional alignments (i.e., the new technology team aligns with information systems, the customer support team aligns with customer support, and so on.)
- **The Virtual Organizations**—The most popular structure today due to recent coverage of this topic.[2] Virtual organizations are a modified form of the horizontal organization. According to Davidow and Malone, the virtual corporation is described as "formerly

---

[2] Ibid, page 5.

well-defined structures beginning to lose
their edges, seemingly permanent things
start to continuously change, and
products and services adapting to match
our [consumers] desires." Although
difficult to comprehend in the context of
today's well-defined organizations, we
can already see the beginnings of virtual
organizations in those companies that
adopt a recombinant structure where
resources can be quickly pulled together
to solve a particular internal or external
problem. However, many of the
participants in these teams are still
functionally oriented. The benefit of
virtuality is the reduction of
organizational response time. The liability
is the cultural impediment created
whenever an organization adopts any
structure; namely, that markets change
faster than most organizational cultures
are able to respond with similar change.
This is the fundamental reason why
paradigmatic leaders, those who break
the mold, have historically come from the
ranks of small start-up companies and not
from existing industry leaders.
- **The Perpetual Organization**—The only
  constant structure that will survive
  change is one that never stops changing.
  A perpetual organization can take the
  form of any structure, based on the
  market demands at the moment. How do
  we break free of the cultural impediments
  that intrinsically bound all organizational
  structures and create a perpetual
  organization? The key lies in understanding how an organization ultimately

changes its form from vertical to horizontal to virtual. In almost every successful case it is the imposition of a subjective crisis, such as a CEO who anticipates a market trend, a large customer who demands change from a supplier, the expectation or realization of diminishing profits, or other quantifiable indicators of change. Notice, however, that I said "every successful case." The streets are littered with the remnants of organizations who realized their predicament long after any change could be instituted. Planning for change is much like blocking a penalty shot in soccer—if you wait until you know the direction of the ball, you have waited too long.

So how does an organization prepare itself for the unexpected and react to events that have not yet taken place? Let's go back to the visionary CEO, who has a certain amount of subjective discretion over the organizations direction and structure. A visionary CEO may redirect the company at a crucial moment and realize a substantial success by entering a new market. But, as we stated earlier, the number of times an organization can redirect itself can be limited by practical consideration.

The alternative is constant and objective feedback from within the organization, accessible to decision makers such as the CEO and everyone else who participates in the process. This information would offer a constant feedback loop to help make decisions that create a perpetual re-engineering effort.

Workfow provides the ability to obtain timely information about processes. With such metrics an organization can open itself to a dynamic culture that embraces change through a constant stream of successful examples.[3]

Nevertheless, a variety of obstacles remain to flattening an organization and creating a perpetual model of change.

Lost Loopholes

Flattening of the organization has other implications as well. It imposes a new level of control on office workers. Loopholes and hidden inefficiencies become evident as existing processes are analyzed and eliminated through the development of automated workflow rules. These rules must be enforced and carefully monitored for process efficiency. The irony is that some level of flexibility and an occasional loophole allows for the introduction of creativity into a process. Few would argue that creativity is not a productive element of office systems that support empowered user models such as TQM. If the workflow system is arbitrarily rigid and tightly structured it will inhibit the ability of workers to participate and contribute to dynamic and spontaneous communications.

More Information than Ever

In addition, these same workers will face enormously increased amounts of information, to the point of overload. Since the natural ebb and tide

---

[3] Again we should emphasize that these metrics are intended to objectively overview and change processes not to scrutinize the workers. Re-structuring the way work is balanced across a group of workers, changing the serialization of a process to a parallel structure, redirecting work to individuals with the correct core-competency, or organizing teams based on a combination of core-competencies required for a specific problem resolution are all examples of using process-focused metrics.

of work will be regulated and transfer times minimized or eliminated entirely, the worker's queue will be constantly demanding attention. The more efficient the business, the greater the likelihood that the user will be the bottleneck. It is the classic man/machine interface problem. The workflow system must be coordinated with the capabilities and the education of the worker in mind. That may mean increasing the worker's understanding of the process through education or tempering management's zeal for productivity in favor of models that evolve with the workers.

Information and its ownership means control and security.

The Proprietary Nature of Knowledge

Finally, if all else is ignored, the worker will lose what is perhaps the most valuable aspect of office information, the proprietary interest in the information and the process that created it. Office tasks and documents are closely held by workers as proprietary material, evidenced by the preponderance of filing cabinets in offices and, for those who have made the transition to electronic documents, the size of disk drives on their desktop machines.

Information and its ownership means control and security. Careers and livelihoods are determined by specialization of knowledge and availability of information. Take this away from workers and they lose interest in the product, service, or task at hand. It is no different on the assembly line. Workers enjoy a sense of pride and benefit from their ownership interest in their knowledge. It is in fact no different than the sense of pride that factories have tried to instill in their line workers by requiring them to stamp their signatures onto completed automobiles, a practice common among many luxury car manufacturers.

Knowledge, whether embodied in paper or electronic form, becomes a linchpin of job security.

Although each of these obstacles can be overcome with some forethought, they are a natural part of the evolution of a workflow application. Some of these can be anticipated, some cannot. An approach that stresses education, an emphasis on the business, an appreciation of the cultural issues involved, and an incremental approach to implementation will provide the greatest likelihood of success and user acceptance of workflow.

Measuring Success

Workflow is a necessary part of re-engineering. But the two terms are used interchangeably far too often. Although there are commonalties shared by the two and synergy of each for the other, there are also some substantial differences. Mainly, these revolve around the relationship of sponsorship to the measurement of process efficiency and payback. This relationship is simple: If a high-level sponsorship (such as that of a CEO or chairperson) does not exist, a bullet-proof case must be made for the payback and improvement in process efficiency. If a CEO is committed to re-engineering, all this is moot. The case study of Connecticut Mutual, discussed later in this chapter in the section entitled *Two Case Studies of Crisis Re-engineering*, is an excellent example of sponsorship being present at the highest level of the organization.

Such sponsorship is hard to come by, as it involves a high level of risk. Surveys of organizations using workflow have repeatedly shown that only 14% of all workflow purchases involve a formal and successful cost justification.[4] What about the remaining instances? Can workflow be justified without demonstrated cost benefit or sponsorship? Yes, but an initial implementation may be the only method to adequately measure the effectiveness of a business process.

To go even further, workflow may be the only timely method of assessing business process efficiency! At first, that statement may sound like a bit of a stretch. What about all of the systems in place to provide managers with reports detailing a

---

[4] 1993 Delphi survey of 400 purchasers of workflow technology.

variety of volumetrics, such as call volumes, throughput, transaction rates, and the rest? Don't these provide sufficient data to make sound decisions? Perhaps, but only if the data they contain is assimilated in sufficient proximity to the actual events that caused the problem. In practice, this type of data is often woefully outdated. A management report that is produced monthly is not only a month out-of-date but typically a month and a half to two months out of date by the time

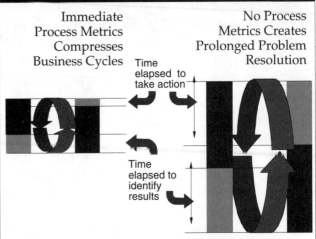

The Control Room Metaphors Provide Metrics for Real Time Problem Resolution

Immediate Process Metrics Compresses Business Cycles

No Process Metrics Creates Prolonged Problem Resolution

Time elapsed to take action

Time elapsed to identify results

Problems are immediately evident when the control room monitors business cycles in real time. Action taken to resolve a problem is immediate as are the results of the action on the business process. The business process can be changed based on these ongoing metrics. No lag time means no guessing as to the effectiveness of problem solving measures. Total cycle time also goes down as a result.

As problems occur, there is a natural delay in identifying them and a corresponding delay in addressing them. The result is that action taken to resolve a problem may not have a direct influence on the original problem and almost no influence on its resolution. The time required to determine the effect of action taken is also delayed. The result is a longer business cycle and no metrics by which to improve the business process.

data is analyzed and distributed. There is an inherent lack of synchronization in management reporting systems. They report on the symptoms of events that have happened in the past, but managers have to solve them in the present. That presents an almost irreconcilable anachronism. This is where the *workflow control room* metaphor offers a new means of associating problems with their true underlying causes.

The Workflow Control Room

The control room is an ideal metaphor for workflow used in this justification context. Control rooms, such as those used in a nuclear power plant to monitor vital functions, provide instant access to information about the myriad functions of the reactor in order to give a plant manager ample opportunity to address a potential problem before it becomes a crisis. Many re-engineering efforts today are undertaken in response to a crisis. But even in a visionary organization, the problem remains that the positive or negative impact of any measurement can be assessed only by evaluating it against prior process measurements. This adheres to the benchmark principle already described,[5] but if we wait for results to indicate the effectiveness of a process change, we will be perpetually behind the response curve or responding outside the optimal time frame for the action to have any effect on the problem.

---

[5] The benchmark principle, discussed at the beginning of this text, states that the efficiency of a process can only be measured in relation to an existing or anticipated benchmark.

Consider the following problem. The customer support organization of a major software provider notices an increase in complaints about responsiveness to support inquiries. The organization ramps up with permanent staff, but due to the length of the interview and training process, customer complaints increase steadily. After the new trainees are in place, complaints decrease. It would seem that the department has addressed the problem. Six months later the same event occurs and the same course of action is taken with the same apparent resolution. This continues until the support function of the organization has expanded to three times its original size. The growth is attributed to increased business volume and the cost of doing business as a software vendor. A closer analysis, however, revealed that customer support representatives were spending less time than ever closing calls and more time trying to coordinate specialized routing of calls to the appropriate analyst.

During a period of downsizing, the support department is one of the first to be targeted due to its size. Workflow software is brought in to enable the effective routing of information to specialists. As a result of the workflow software, it becomes apparent that customers are calling in several times to ask the same question of several analysts because individual analysts do not have adequate knowledge of the intricacies of the integrated software system to handle more than a sliver of a customer's problem. Redundancy abounds in this scenario as each specialist repeats much of the work and research done by his predecessor.

The solution becomes obvious: Generalists take the place of many specialists in the department, leaving behind a handful of individuals able to respond to very specific issues under the guidance of the generalists, who act as conductors of a symphony orchestra.

The actual problem in this scenario was never clear because cause and effect were not clear. The fact that complaints tapered off after additional hiring was not in response to the new analysts but to the fact that the new staffers started as generalists who took responsibility for gathering all of the information needed to answer a customer problem. Once they became specialized, however, the problem resurfaced causing the cyclical pattern in customer complaints.

The control room metaphor would have highlighted the short-term peaks in call volume and the distribution of calls, across analysts, in a real-time monitor. That monitor would have provided instant information to justify changes in staffing, workloads, information flows, and the re-engineering of the support function. The downsizing imperative and the application of workflow made evident a problem that would otherwise have been obscured and ignored. Although downsizing is not a pleasant experience for any organization, it is often the only way to uncover problems of this type—problems that often end up labeled as, a cost of doing business.

The bottom line? There are organizations with sponsors willing to undertake a re-engineering effort. These sponsors have the budget and the clout to enforce change. The metrics for change, however, do not often exist. Even where metrics exist, they may be associated incorrectly with an erroneous cause-and-effect relationship. When implemented to automate existing processes, workflow can establish baseline metrics and long-term monitors for the efficiency of business processes. These metrics can then be used as a stable, sound, and reliable fuel for long-term re-engineering efforts.

The Effect of Workflow on Communication: P2P Communication

A profound effect of workflow has to do with the fact that it will significantly change the nature of communication. At the heart of this is *asynchronous communication*; a means of enabling communication that allows business processes to bridge time and distance. Think now of our earlier discussion about the analogy of the expanding balloon and the importance of fostering communication to increase intimacy with adjacent steps in a process.

The concept of asynchronous communication is a simple yet powerful one. Standard interpersonal communications occur in a *synchronous* mode. That means that when two people conduct a discussion they are able to communicate at the same time, and possibly in the same place. The benefits often attributed to synchronous communication have to do with the ability to address issues as they arise without delay. Asynchronous communications, on the other hand, occur in series without interaction and interruption. Although contemporary forms of asynchronous communications may come to mind, such as E-mail and voicemail, asynchronous communication is as old as mankind. Etchings on cave walls, and every form of written communication since represents asynchronous communications.

Asynchronous communications are often associated with delays, as each party is always waiting for the other before continuing the communication. This occurs when a task involving several messages (communications) requires that a response correspond to each message. This can be especially problematic if the parties communicating are separated by several time zones. The following example shows the effect this will have on a simple task involving several communications.

When Boston sends a message to Hong Kong at 9:00 A.M. (10:00 P.M. Hong Kong) Hong Kong will not receive the message until the beginning of their day (10:00 P.M. Boston). Although Hong Kong may work all day on the task and make significant progress on it, Boston will not be aware of the progress until the task is complete and they receive a communication from Hong Kong. If Hong Kong sends its response on the day after the message was received from Boston, the total delay, from Boston's perspective, is at least two days. The total task time, however, is only half of the total elapsed business cycle time.

That may seem like an insurmountable problem. After all, no technology can alter the arrow of time. But we can eliminate the fundamental obstacle in synchronous communication—the concurrency of human communication. With an automated workflow system, asynchronous communication can continue even though the people are not synchronized. Instead we synchronize people with processes and tasks. In our example of Boston and Hong Kong, Boston would be able to communicate with the task at any time.

This is a fundamental change in the nature of communication, as we are no longer communicating with people but with the process itself. Note that communication in our definition requires that there be real-time interaction between the two parties. (The parties, in this case, could be any combination of human and process.) Without an automated workflow system this can be accomplished only if two individuals are both available at the same time. Let's refer to this as *P2P (Person-to-Person) communication*. Workflow captures the process in such a way that this communication can continue at

...no technology can alter the arrow of time. But we can elimate the fundamental obstacle in synchronous communication— the concurrency of human communication.

any time. We can continue using P2P to refer to this as well. Here either "P" can be a Process or a Person, which adds another wrinkle to asynchronous communication—Process-to-Process. Nothing prohibits situations where two information agents, equipped with the rules by which to make a decision, can execute a Process-to-Process communication and then proceed to the next task. In this way people could be involved in some tasks and then represented by surrogates, or "agents," in other process. This fundamental shift from person to process is one of the most significant aspects of workflow automation.

Re-engineering and Workflow

The most-often-mentioned obstacle to workflow's adoption is that of cultural resistance to change.[6] Of these mentions, almost half also cited re-engineering as part of the cultural obstacle. In many cases the two issues, workflow and re-engineering, are one and the same. Organizational culture is resistant to change, and sponsorship of change is unavailable. Unlike re-engineering, however, workflow does not require radical change and exceptional sponsorship in the organization. In fact, re-engineering is facilitated by the use of an incremental or pilot implementation of workflow technology. This provides a tremendous amount of experience to the implementing organization. It also shows immediate results, which provide a leverage point in justifying future enterprise applications. Initially workflow and re-engineering are symbiotic, in that one feeds off of the success of the other. As the sponsorship of re-engineering and the scope of a workflow application rise within the organization, the two become synonymous.

Re-engineering is often incorrectly thought of as a radical process of decimating and then rebuilding information systems.

Using Workflow as a Metric for the Re-engineering Process

In the territorial wars fought over technology, each group of disciples professes that its approach is the panacea for ailing organizations crumbling under the weight of antiquated business processes and technology. Imaging, EDI, E-mail, workflow, and groupware have each become a fiefdom of technology bigotry. In this quagmire, many buyers are turning to the one option that seems to be the only hope for a solution—re-engineering.

Imaging, EDI, E-mail, workflow, and groupware have each become a fiefdom of technology bigotry.

---

[6] 1994 Delphi study of 400 users and evaluations of workflow.

The popular view of re-engineering as the obliteration of existing process and systems has its own set of issues and risks. First, the level of sponsorship required to cause such drastic change to occur within large organizations seldom exists. Second, even sponsorship requires some direction. Where should re-engineering begin? Which business processes and information systems are least efficient? What is the measure of efficiency?

Without answers to these questions, any re-engineering exercise is no more than a roll of the dice. Workflow can provide these answers and in the process establish the foundation for a sound re-engineering effort. But to do this, workflow must be used as more than an automation tool. It must also provide the analytical and reporting tools that help change-managers better understand their organization's business processes. In this light, workflow becomes an overall discipline for restructuring information systems, not another competing technology.

Although workflow is part methodology, it is not synonymous with re-engineering. The two are, as we have seen, complementary methods. Re-engineering proposes the obliteration of existing work processes and the establishment of new business methods based upon new sets of assumptions regarding desired business objectives. It is a comprehensive approach to redefining the organization. Workflow is the analysis, compression, and automation of information-based process models that make up a business. In short, workflow provides the metrics for re-engineering.

Although workflow appears to represent only one component of total re-engineering, no re-engineering project should proceed without the use of workflow, at the very least as an analytical tool. The reason? How can you undertake a redefinition of an organization if there is no benchmark against which to measure the efficiency of its business processes? As we will see, tools such as Time-based Analysis offer the means by which to establish these benchmarks within your organization.

But measurement alone will not foster change without the application of some form of sponsorship to the re-engineering effort. The specific form of sponsorship, and the motivation behind it determines one of three methods of re-engineering.

The Dimensions of Re-engineering

Re-engineering is not a one-dimensional model. Three models of re-engineering are commonly used: the life cycle model, the crisis model, and the goal-oriented model. Although each method may have its place in an organization, they should not be regarded as interchangeable. An understanding of each approach is necessary to determine which is appropriate for your organization.

Life cycle re-engineering results from a strategic initiative to constantly re-evaluate existing processes. Change is incremental and basic processes remain essentially intact, modified slightly to accommodate new requirements. Modification is usually aligned with the critical success factors of the organization. Automation is often applied, but the primary emphasis is on enhancing and streamlining the process. The sponsorship must be cultural, not mandated.

Crisis re-engineering is a response to systems crumbling under the weight of user demands or organizational pressure. For example, a service firm with a cumbersome billing process had a high percentage of fees earned not billed. The prolonged billing cycle resulted in a negative cash flow and led

The three methods of re-engineering are: Crisis, Life-cycle, and Goal-oriented.

directly to a crisis re-engineering of the billing process with the objective of speeding client invoicing. Sponsorship is not necessary since change must occur regardless of the solution chosen. Re-education and technology evaluation are used rather than a formal re-engineering methodology.

Goal-oriented re-engineering has defined objectives that may differ substantially from the objectives in place when the system was first developed. This is a deliberate attempt to bring an existing process in line with business objectives. Business processes are totally redesigned with new goals and objectives in mind. Sponsorship is inherent. The sponsors must have the clout to drive the application through the organization, often without cost justification. A re-engineering methodology such as a Time-based Analysis tool is almost always used in this case, since the emphasis is on a strategic, long-term implementation with an extended payback.

Life cycle re-engineering is the least disruptive approach because it fosters an opportunistic attitude towards change and an incremental, ongoing approach to modifying information systems. Customization is minimal because the change is spread out over time. The risk is that significant inefficiencies may be overlooked because of the effort required to re-engineer them. Also, it can be used only if organizational critical success factors (CSFs) are well-stated and supported among management and users. Workflow is best applied in this type of a re-engineering mode, because the workflow will provide an ongoing measure by which to constantly modify and refine the business processes.

A good workflow tool generates on-line reports that offer insight to the workloads, bottlenecks, resource allocation, throughput, productivity, and the overall business cycle. By analyzing these, immediate decisions can be made to alter a process by re-allocating resources, changing task relationships, eliminating redundancy, or altering priorities of work. The result is a highly adaptive and responsive organization—not unlike the model used to retool factory assembly lines for the purpose of mass customization.

The approach used most often, crisis re-engineering, is thrust upon an organization with no other choice. The decision is not, "Should we re-engineer?" but rather, "Can we afford not to?" Organizations that rely on crisis re-engineering are doomed to encounter a re-engineering crises regularly. The prevailing attitude in these organizations is "well enough is best left alone." Entropy increases regularly until action must be taken. Since the re-engineering is forced by a particular scenario, the analysis is often biased and rushed. This method has become the norm for re-engineering.

Organizations that rely on crisis re-engineering are doomed to encounter re-engineering crises regularly.

During the industrial revolution and the decades leading up to the present day, technology obsolescence occurred at regular, planned intervals measured in decades. In that environment, it was possible to retool and re-engineer once or twice each decade. Today, technology is obsolete in months, not years. Re-engineering can no longer be an interval activity—if it is, the re-engineering will always be too late to do any good. Ask the fundamental question, "Does my organization budget a certain amount of money each year, for re-engineering?" Without that type of commitment—let's say a few hundred dollars for each technology-enabled desktop—re-engineering will always be overdue by the time its mandate arrives.

The most enduring approach, goal-oriented re-engineering, requires high-level corporate sponsorship. It is driven by the anticipation of future benefits resulting from re-engineering rather than the cost of an old or inefficient solution, although both should be considered in the cost justification. The primary difference between this approach and the others is the emphasis on business processes rather than technology.

Workflow is compatible with goal-oriented re-engineering so long as sponsors demonstrate a quantifiable measure of business process improvement. Workflow assists in re-engineering through its ability to monitor and report on changing business cycles. Unfortunately, technology is often separated from business process redesign in most re-engineering approaches. For many, that has meant that workflow should be left out of the re-engineering effort, in order to avoid compromising business goals. That need not be the case.

As workflow does not impose a specific technology tool set of its own for the actual performance of work—office automation tools, such as word processing, database, spreadsheets, E-mail, and EDI should all work within the workflow environment—it does not alter, impede, or compromise the re-engineering effort.

It is safe to say that, not only can multiple re-engineering approaches exist in a single organization, but there may be elements of any two or all three approaches applied simultaneously to a single re-engineering effort. In practice, the most likely case is that all three will come into play. For example, a crisis may initiate the re-engineering effort; management then identifies a long-term goal that provides a context for the initial project and a series of subsequent re-engineering efforts intended to achieve a strategic vision; finally, the vision itself is tested repeatedly in a life cycle approach that continuously validates the vision against the reality.

Weight loss is a striking analogy. Many people struggle their entire lives with weight gain and loss—an endless roller coaster. On occasion, however, a crisis will occur, perhaps a physician's warning of health risks, an actual health crisis, or something as subtle as seeing one's self in a family photo. The impetus, is strong enough to cause an individual to take action. In some cases that action and a bit of success turns into a vision of being on a regimen. That vision needs to be reinforced with success, since it is not of itself a pleasing experience—results are what drives the vision at this stage. Over time, however, the only successful weight loss program is that which not only reinforces the vision of a regimen, but that of a person and a lifestyle. At that stage the individual's every decision and future action will have a bearing, not on weight but instead, on lifestyle. By this time, there is no regimen, only an ongoing process of checks and balances that act as the measure and countermeasure of success.

Now if your organization can, or has been able to skip past the crisis and goal-oriented stages of re-engineering, you are simply that much further along. Try some basic tests first to see where you now stand:

For the Lifecycle Re-engineering Organization

What are the critical success factors (CSFs) of your organization? (You should be able to state these no matter what your position, but a little bit of digging is allowed. A lot of digging means that the CSFs, if they exist, are token gestures, not cultural mandates.)

How much do you budget a year for re-engineering?

Do you even have a budget line item that mentions re-engineering?

Do you have a title in your organization that contains the term re-engineering (or a derivative)?

For the Goal-Oriented Re-engineering Organization

Do you have a clear vision of the new organization or process (The Goal) you are trying to create?

What is the regimen you will follow to achieve this vision?

Is there a sponsor behind the vision? (Notice we did not ask this for the life cycle organization, because sponsors will change on an ongoing basis in life cycle re-engineering.)

Does the sponsor control the budget or control the person who has the budget?

The following four case studies demonstrate the application of each re-engineering model. In each one, elements of crisis, life-cycle, and goal-oriented re-engineering can be found. Each has been categorized by the *predominant* model that was the force and impetus for change.

Canadian Broadcasting Corporation (CBC) Management views its re-engineering effort as a living plan and has continually re-tested its assumptions against its goals.

A Case Study of Life-cycle Re-engineering: Canadian Broadcasting

Many organizations view re-engineering as a radical extrication of old business models. Canadian Broadcasting Corporation (CBC) management views its re-engineering effort as a living plan and has continually re-tested its assumptions against its goals. The result is a solid re-engineering case model for organizations that are interested in changing business processes without encountering radical and disruptive change.

This case study looks at a class of business re-engineering common to many organizations— financial systems. The CBC, Canada's national broadcast service, used workflow and document imaging to consolidate the backroom payables functions of eleven regional offices and, in the process, achieved a doubling of productivity. The project is expected to pay for itself in less than three years. In the course of the project, the CBC faced the challenges of centralizing well-established regional systems and balancing change with union contracts. The accounts payable project described here will become the model within the CBC for future re-engineering efforts.

Caught in a Costly Paper Chase

The CBC produces and transmits television and radio programming in English and French, as well as aboriginal and northern languages. The CBC combines centralized and decentralized management structures across all eleven provinces and both territories of Canada, with production and/or broadcasting and administrative units in each. In 1992, when Finance and Administration initiated an accounts payable re-engineering initiative, the CBC had already undergone some downsizing, eliminating three of its regional offices through consolidations.

One of the principle reasons for the CBC's re-engineering initiative was the processing of invoices, which had become highly paper intensive. Costs would skyrocket whenever it was necessary to stop a process to investigate errors, check whether payments had been made, or research discrepancies. Although not of crisis proportions, the process costs were steadily increasing. The CBC turned to workflow and imaging hoping to significantly alter the amount of paper handling required for accounts payable.

High Costs and Low Productivity Levels
Made Compelling Case for Change

CBC line managers began their streamlining efforts by researching a number of large companies that had undertaken re-engineering, including Chemical Bank, which had re-engineered its accounts payable, and Amtrak, which had re-engineered order processing operations. They found CBC's accounts payable productivity, at 4,000 invoices per person annually, was far below the levels of peer companies, which were processing in excess of 10,000 invoices per person annually. There was an obvious discrepancy with the benchmark and CBC's productivity. CBC noted that in each case where organizations used automated workflows they had consolidated operations, used strict procedures to reduce exception handling, and eliminated paper processes. Tours of these operations convinced CBC management that it could increase accounts payable throughput to 10,000 invoices per person annually, or even higher levels, representing more than a 250% increase in productivity. The benchmarking approach used by CBC is fairly standard. Other business models of direct competitors or like-industry leaders are used to measure the effectiveness of a business processes in a similar industry context. But, as we will see in the Connecticut Mutual case study, this is not always the case. Stepping outside of your own industry may yield equally valuable benchmarks.

Based on CBC's research, a justification was formed for the re-engineering project that resulted in adequate start-up funding and commitment at all levels of CBC management. Having seen it work, they understood the significant internal issues, staffing, union obligations, organizational politics, and the re-engineering approach. Beyond that, the project would give the CBC an experience base for re-engineering other functions such as travel accounting, credit and collection, and local payrolls. By using the same approach for these solutions, the CBC would optimize the return on its initial investments. To establish a firm metric for success, a payback period of less than thirty months was targeted for this specific workflow solution.

The implementation strategy focused on change in manageable increments (one regional office at a time) and began with the most complex aspect, integrating imaging technology with the current financial systems.

The CBC made every effort to minimize customization in its implementation of the workflow package to keep costs down and still meet the needs of the organization. The system uses advanced imaging and workflow techniques and can be quickly tailored with reduced risk. It was hoped that this approach would provide a track record of tactical successes upon which to justify and build future workflow applications.

Re-engineering: From Eleven Regional Offices to One National Payment Center

One of the key ingredients of the CBC's success in implementing workflow was its emphasis on looking beyond the narrow scope of just the paper invoicing problems. Document flows and system functionality were closely scrutinized, including reviews of credit card use, arrangements with suppliers for direct purchases, changes in internal processing methods, expansion of PC-based purchasing applications, reduction in small value purchase orders, changes in processing small value invoices, and general streamlining of all procurement processes. This led to the restructuring of certain value chain activities in ways that may have been obscured with the implementation of a point solution. This is clearly evidence pointing to the possible marriage of workflow and re-engineering.

In effect, using a few generalists rather than numerous specialists, almost always acts to shorten and simplify a business cycle.

For example, a portion of the workflow planning required realigning document streams with an eye to eliminating unnecessary paper handling. In one such case direct invoices (those without purchase orders) would be sent to managers after passing through layers of local administrators, which inevitably slowed the process. By shifting the focal point of the re-engineering from the finite scope of the local operation, it became clear that the establishment of a National Payment Center in Ottawa, as part of the Corporate Finance Group, would make it possible to use existing resources for this purpose. (Ottawa already housed the CBC's Finance and Administration Group and this would allow the merger of accounts payable activities with other corporate finance services without adding layers of new management.) This would eliminate the need for the local administrators' and the managers' involvement in the process. The center would process all invoices and payments as well as handle corporate payroll activities, local payroll, and travel settlements, and in time the payment center would take on other consolidated processing functions.

This aspect of the workflow would reinforce the importance of having a centralized control function to oversee and coordinate the workflow process without adding management for this purpose—a key benefit of many workflow applications. The result is a single point of ownership for several process steps. In effect, using a few generalists rather than numerous specialists, almost always acts to shorten and simplify a business cycle.

Even with Workflow, Paper Has a Place

The National Payment Center, now in its first phase, handles all accounts payable operations once manager approvals or verification of receipt of goods is noted. Because the CBC's normal business practice is to mail invoices to the person who placed the order, regional locations still receive (paper) documents for approval. These are then forwarded to the National Payment Center via mail.

It is interesting to note that the CBC planned its workflow system with the expectation that paper would exist for a time. Workflow and document imaging, however, have eliminated the need to handle physical paper once an invoice has been scanned into the system. This type of paper elimination is often referred to as a *point-of-entry/ point-of-access*[7] system. Paper is eliminated during a substantial part of its lifecycle, typically starting at the point-of-entry into an in-house workflow system. From that point on, the document is retrieved from point-of-access workstations that have the ability to view all information pertaining to the document and transaction. Although this does not entirely eliminate paper it optimizes the business cycle during the document's electronic lifecycle.

[7] Koulopoulos and Frappaolo. *Electronic Document Management Systems*. McGraw-Hill, 1995.

Implementation Goals: Avoid Re-entry, Minimize Risk, and Control Costs

A primary goal of the CBC project was to integrate workflow and imaging with legacy systems already in place. This served to reduce development costs, eliminate data re-entry, and minimize the risks inherent in an "all new" system. In linking the workflow system at the Ottawa Payment Center with the mainframe in Toronto (the site of the existing accounts payable system), the CBC discovered features within the accounts payable system that could be used to achieve centralization and still allow local check production for critical situations, local access to vendor files for customer service inquiries, and local cash management information. These were used extensively. Because the accounts payable system links directly to the general ledger, no additional links to the workflow system were required.

As documents are scanned, the system encodes them with an electronic ID and, as indexing takes place, an electronic file is built for each. The system imprints the same ID number on the paper to make the audit trail complete. Image files support the field-oriented data recorded in accounts payable, which, in turn, feeds general ledger and financial reports. Because the invoice number is fed through to the general ledger, remote managers can query the National Payment Center for detail and receive copies of the invoice by fax. On-line inquiry to accounts payable and check printing is also possible. This closes the information loop, much better than what was possible with CBC's paper-based systems.

Enriched Job Functions

At the CBC, re-engineering resulted in a significant reduction in staff across the system. This was accomplished in part by establishing a new policy that specified the use of temporary personnel to replace anyone who retired, left the corporation, or moved into another area. Internal transfers were encouraged, and retraining was offered with the cooperation of other departments. By instituting such an array of creative options for downsizing the CBC effort was not perceived as one whose intent was reduction of staff. As CBC was not in a crisis mode, it could afford the luxury of phasing in downsizing over a longer period of time.

Among the key factors to successful workflow, which CBC uncovered, was the selection of the management group and senior clerical staff assigned to the project team. The manager, who was hired to start the project, hired two supervisors who in turn were involved in the selection of the remaining staff. Bringing in outsiders can work both ways: to add a fresh perspective or to alienate workers. In this case new management spent considerable time educating staff prior to implementation.

Training the clerical staff took a month of effort and a significant amount of hand holding before clerks achieved proficiency. Previously, the clerical staff had used only dumb terminals. The workflow system uses windows and on-screen techniques, creating a paperless office once documents are scanned. Clerks use twenty-inch dual page screens for viewing imaged documents.

Finance provided training for managers, developed coding books, and designed the electronic mapping of data to the screens that linked the document system to accounts payable. When the implementation went live, the manager and

As CBC was not in a crisis mode, it could afford the luxury of phasing in downsizing over a longer period of time.

users were able to redesign and fine-tune the screens
as necessary—thus encouraging direct end user
involvement in the final system definition.

Assessment: Productivity Levels on the Rise; Staffing Cut by Half

As of June, 1993, the CBC moved approximately
60% of its corporate volume to the national system
(six months after implementation began). When
the accounts payable system is fully implemented,
the number of invoices processed per person is
expected to be closer to the goal level of 10,000 and
the CBC expects staffing reductions of 50%. The
remaining regional offices will be added after
Finance has consolidated travel and local payrolls.
In smaller regional offices, a single individual often
handles all of these functions (accounts payable,
travel, and local payroll functions) making it
difficult to reduce "a part of a person" when only
one of his or her functions has been centralized.

The CBC attributes much of the success of its re-
engineering to the level of communication between
planners, implementers, and the people who
actually use the system. Managers and supervisors
were very involved in every aspect of design, as
well as the overall implementation. The system
became *theirs*. That was one of the early goals of the
project.

Currently, CBC management is looking into
implementing a variety of feedback mechanisms
for the processing clerks. Worker satisfaction has
remained high despite the ongoing challenge of
higher productivity goals because they sense they
are on the leading edge of the re-engineering effort,
without risk of radical upheaval. The re-engineering
effort recombined and enriched job functions; clerks
are working with newer technologies, applications,

and methods. Individuals contribute ideas for changes and improvements.

The CBC's implementation of workflow demonstrates that an incremental approach to workflow is not only possible but in many cases the most appropriate approach in large distributed organizations with user communities that may be resistant to massive change. CBC management was astute enough to use an initial benchmark. Of even greater importance, however, was the fact that the CBC's workflow implementation created its own benchmark within the CBC. This first-hand experience and the education provided is the most compelling reason for users and sponsors to continue their investment in re-engineering remaining business processes using a life-cycle approach.

First-hand experience and the education provided is the most compelling reason for users and sponsors to continue their investment in re-engineering.

Hospital Revamps Patient Records Using On-line Imaging and Workflow

Workflow projects often fall short of expectations because of a lack of committed sponsorship or because of subtle resistance from users. A classic strategy for overcoming these obstacles has been to identify a pocket of sponsorship and start with a small test case that will demonstrate the technology and its benefits for the rest of the enterprise.

St. Mary's Hospital took a dramatically different approach. It targeted emergency services, one of its most visible departments, as its test case—even though there were other areas in which it would have been able to have demonstrated a greater payback sooner. Equally remarkable, this project requires clinical users who were unfamiliar with the use of the computer as an integral work tool to give up paper in favor of totally electronic record keeping.

Workflow projects often fall short of expectations because of a lack of committed sponsorship or because of subtle resistance from users.

This case study documents the planning St. Mary's went through in preparing for its on-line patient records system. The Emergency Records phase of the implementation went live November 29, 1993.

St. Mary's Hospital of Grand Junction, Colorado, is the only hospital offering air ambulance service, a dialysis unit, a Level 2 nursery, and certified trauma and oncology centers in a referral range that spans from Denver to Salt Lake City. The hospital had not changed patient records management practices dramatically since its founding in 1895. St. Mary's held to tight financial management practices. The costs associated with patient information had serious implications to its financial picture. (American Health Information Management Association, AHIMA, estimates project health care facilities spend 40 to 60 cents of every dollar on paperwork. A recent article in *The New England Journal of Medicine* called 25% excessive.)

With Continuous Quality Improvement as a corporate goal, St. Mary's initiated a review of its information handling practices to reduce redundancies and rework areas where it could stabilize and improve its systems. Records management offered a well-defined area in which management believed the application of image processing and workflow software could reduce administrative costs and improve operational efficiency without jeopardizing quality care to the 11,250 inpatients and 61,900 outpatients who use the hospital each year.

A $1.2 Million Project that Will Pay for Itself Over Five Years

There was no lack of commitment to St. Mary's project. Hospital management, including Finance, MIS, and the Health Records Department, were all firmly behind the effort. The Director of Health Records Information Services (HRIS) developed the business plan for the restructuring of patient care records to replace time consuming, obsolete manual and microfilm methods. St. Mary's Continuous Quality Program compelled management to identify ways to introduce quality improvements throughout the hospital while maintaining and reinforcing its patient focus. In restructuring health information services, St. Mary's wanted to be able to give multiple caregivers simultaneous access to information. The workflow solution was crafted to eliminate the need to copy and courier records, to reduce paperwork, streamline office procedures, minimize mistakes, and improve patient care.

With these goals clearly outlined and the plan for a demonstrable return on investment, the sponsors were able to successfully route the business plan through an elaborate five-tier approval cycle. Key departments (Health Records, Patient Accounting, Finance) identified how this program would help them redesign processes and committed to resulting reductions in staff. MIS and HRIS committed to decreased costs in records storage, paper, and microfiche. Finance committed to reductions in the accounts receivable cycle and a more balanced cash flow. The final analysis budgeted $1.2 million for the installation. The project was expected to pay for itself over five years.

Emergency Department Is Test Case

The project team's ultimate goal was to implement
workflow and imaging throughout the hospital.
Initially, however, workflow and records imaging
was applied to the patient care component. This
was an interesting choice and crucial to the ultimate
success of the project. Beginning with patient care
in the Emergency Department, a highly visible
area, would require greater discipline and show a
slower payback than beginning in Finance. This
challenge was balanced by the belief that this was
the best place to show the effectiveness of having
quick access to information from the moment a
patient arrives, reinforcing the value of patient-
centered care, a clearly defined corporate goal.
Eventually, workflow would touch all aspects of
hospital operation.

MIS was actively involved in all phases of the
project to ensure that the dual perspectives of
technology and records management were
represented. HRIS personnel worked closely with
MIS in areas where patient records were concerned.
Overall, the responsibility for orchestrating a
hospital-wide approach to imaging, carrying the
technology beyond health records to patient
accounting, accounts payable, and human
resources, ended-up in the lap of MIS. MIS was best
suited to represent the interests of all parties whose
infrastructure may have needed to tie into the
system. This decision also demonstrated the trust
placed in MIS by the organization, to best secure its
investment in creating enterprise processes. Given
the rate of change of the volatile workflow market,
that turned out to be crucial.

First Step: Outline the Work Process

In the first year, the Emergency Department projected that it would generate about 180,000 documents, including demographic information, physicians reports, lab and radiology results, and nursing documentation. As a first step, the department created flow charts of the existing manual process, which involved 32 steps.[8] Analysis revealed that the average retrieval time for a document was between eight and fifteen minutes, but it was not unusual for some records to require more than an hour to retrieve, such as documents involving microfilm documentation of multiple visits. All records were hand carried and, of course, locating a document often required a lengthy walk. Accessibility to current information suffered also. After a visit, the physician would dictate notes to the emergency department record. The following day, a clerk would sort documents from the previous day, remove carbon copies, re-assemble the material, and then determine if the records were complete.

The planned workflow and imaging system reduced the overall process to twenty steps. Emergency caregivers are now able to retrieve patient records on-line in a matter of seconds. The optical-disk-based system gives doctors two choices for dictation: a Kurzweil voice-activated Auto-Report, or notes transcribed from a digital dictation system. The transcription is typed and stored on-line as a data file, along with laboratory and radiology reports using computer output to laser disk (COLD). This information is now also available to any authorized user. If the doctor needs a document, he or she recalls it on the computer screen, rather than sending runners to retrieve it.

---

[8] We will propose a method for creating these flowcharts in the section on the *System Schematic*.

Another step in the automated process is scanning handwritten documents (such as physician's orders) into the imaging system where they are indexed and verified. If the information is correct and complete, it is approved and sent on for automated verification. If not, the document is re-scanned.

Approximately fourteen physicians and twenty nurses and technicians share three desktop workstations located in the administrative desk area of the Emergency Department, where dictation is normally done. Interestingly, St. Mary's made a conscious decision not to provide a printer. Users were then compelled to view information on the computer screen. This ensures the confidentiality of the information. More important, it encourages people to change their work patterns by using electronic records. The goal is not simply to speed up the process, it is to keep information on-line and instantly available, thereby eliminating as much paperwork as possible.

Physicians unfamiliar with Microsoft-Windows® software received about a half-hour of instruction at the time of system orientation. The other emergency department staff members received about two hours of instruction for proficiency. The Health Records Information Services staffers, the people responsible for maintaining the patient records, received about an hour of Windows tutoring followed by two hours of basic image retrieval and access training. A select group responsible for indexing and scanning received an additional four hours of instruction. With this meager amount of training, each user is adequately prepared to participate in the benefits of the system. That is a crucial point, since many technologies are resisted primarily due to the investment of time required to become proficient in their use.

Preparation: A Thorough Review of Documents and Procedures

The heavily paper-based system in existence when the project was started required St. Mary's to evaluate its existing inventory of documents before converting the existing systems to an electronic model. This was easily the most extensive component of the process, and is one that most organizations underestimate. For the year prior to implementation, the Health Records Information staff had been revising documents to eliminate as many unnecessary records as possible and to make sure that the ones they would use complied with the hospital's Forms Revision Policy and Procedures. In order to maintain a life-cycle approach to their new systems, St. Mary's staff review all new forms to verify that they are designed to meet the needs of patient care, not the computer. This is indeed a solution committed to supporting the critical success factors of the organization and not the demonstration of novel technology.

Lessons Learned: Process Redesign is Fundamental, Know Your Users

In assessing the project, MIS believes that the key to success was the time spent in re-engineering hospital processes. As it looks to initiate new projects based on this model, MIS will place even greater emphasis on process redesign. While this was done extensively for the Emergency Department pilot project, attention was split between process redesign and the need to learn the new technology—imaging, workflow, scanners, and software. Experience has shown that the preparatory work done in understanding the processes, designing and standardizing forms, and finding the leverage points is the key to the success of the project.

In reviewing the response of end users, the sponsors were surprised to find the impact of change was somewhat greater than originally anticipated. Because the group selected for the pilot was a clinical department, there had been minimal previous exposure to computers and little experience using desktop technology as an integral work tool. It took a little more training in the use of the desktop workstations and Windows software than would have been necessary if the pilot group were a finance department already familiar with computer technology. MIS relied on unorthodox means to ramp up these end users—computer games. Installing these on workstations turned out to be an easy way of helping users achieve a comfort level with the systems. Playing games helped a number of users achieve proficiency with the desktop and the Windows software and increased their confidence in working with the on-line forms. It also proclaimed the openness with which MIS was approaching the solution. This was a case of true collaboration. Team members were interested in making the systems work, not merely in handing off a solution to end users.

The sponsors were surprised to find the impact of change was somewhat greater than originally anticipated.

The List of Candidates for Workflow Grows

Perhaps the best example of the system's success is the fact that other major areas of the hospital are eager to undergo similar re-engineering efforts. Near term plans call for a tie-in with outpatient centers to provide access to ambulatory patient records. There have been discussions over linking the system to physicians' off-site offices; the first step toward a virtual hospital. MIS is now preparing for the implementation of imaging and workflow in outpatient surgery, another paper-intensive area.

Team meetings are underway involving individuals from all areas involved to flow chart and analyze the steps involved in preparing information for each successive group.

As with so many successful re-engineering efforts, the success at St. Mary's points back to three factors: the charismatic sponsorship of a single individual, in this case hospital president Sister Lynn Casey, a tremendous trust and collaboration between information systems staff and user departments, and the identification of a starting point, emergency care. None of these can be found by formula. No, not even the formulas that will be proposed later in this book. Sponsorship, collaboration, and stated critical success factors can be nurtured, they can even be unearthed with some effort, but they cannot be created from the ether— they must first exist.

Do you have as much to work with in your own organization? If not realize that some successes are smaller than others. Great success requires a great foundation—that formula will not change. A small success, however, may be sufficient proof that it's time for a new foundation.

Summary

We've already seen that many organizations do not prepare a formal cost/benefit analysis when planning a workflow implementation. As the St. Mary's experience indicates, having this information helps to build a case for the introduction of the desired new technology. It is interesting to note, however, that this organization didn't limit itself to the quickest payback, but also took strategic issues into consideration in choosing its pilot application. St. Mary's didn't base its decision strictly on the numbers, but rather invested in the long-term success of the change initiative. Also, the strategy gave St. Mary's the justification to target a highly visible area as the test case. Now that the revised work process is on-line, the cost-benefits analysis serves as a benchmark against which management will measure progress at many junctures. Fine-tuning the system becomes that much easier when everyone involved knows what must be done and where the organization wants to go.

Two Case Studies of Crisis Re-engineering: Connecticut Mutual and Waverly Press

Connecticut Mutual: A Case Study in Successful Workflow

At the outset, management recognized that for re-engineering to be successful, enterprise-wide, there had to be a commitment to common utilities and architecture.

One of the leading role models of workflow in practice is that of Connecticut Mutual Life Insurance. Connecticut Mutual Life Insurance company ranks among the top twenty-four U.S. insurers with assets of more than $12 billion (and more than $90 billion of life insurance in force). Historically, the insurance industry has been characterized by paper-driven, clerical-based, sequential, factory-like workflow. Connecticut Mutual realized these workflows, which are labor-intensive and inefficient, no longer met its service and financial objectives and would in time adversely affect its competitive position in the rapidly changing financial services industry.

Connecticut Mutual's management responded to the challenge with a vengeance. In the fall of 1990, the company embarked on a project to create ONE IMAGE of Connecticut Mutual. The project is noteworthy in its emphasis on change in corporate culture and in operations. It supports goals stated by senior management to radically redesign the financial and organizational thrust of the company to enhance customer service. ONE IMAGE pulls together fragmented systems, eliminating paper, and restructuring and compressing a complex business work process. It is one of the most comprehensive, enterprise-wide implementations of corporate re-engineering in effect today.

The most salient aspect of ONE IMAGE is that it is both corporate vision and management philosophy. At its inception it was endorsed by senior management and enterprise-wide by division heads. It still is. That endorsement is reaffirmed to the entire organization at the beginning of each year.

The Process of Re-engineering

Business re-engineering is a radical rethinking of business processes at a series of levels, not just on the part of the sponsors . A full scale re-engineering effort involves substantial cultural change throughout an organization. At Connecticut Mutual, re-engineering was company-wide, from top to bottom. More than simply implementing a new suite of technology tools, Connecticut Mutual undertook an elaborate self-analysis—looking at itself as an organization, a business, a creator of product, a payer of benefits, and at all the other parts of a complex multi-line company. This radical rethinking led to a series of stages:

Business re-engineering is a radical rethinking of business processes at a series of levels, ...A full scale re-engineering effort involves substantial cultural change.

- A corporate vision of re-engineering—a high level conceptual or pictorial view
- Business area re-engineering—a macro line view of an entire division, such as individual life insurance or financial services

- Workflow views or process re-engineering
- Systems re-engineering, implemented at four levels:
  - graphical
  - navigational
  - hardware
  - software

Connecticut Mutual also identified four areas that it considered fundamental to its business: customers, the workflow, the information, and the technologies it would need to effect an enterprise-wide re-engineering.

Building the client view required all groups to look at customers and producers in detail, then focus all corporate activity from sales to systems on the key service goal of more rapid and thorough customer response. Rather than target parity with other insurance and financial services providers, Connecticut Mutual modeled its customer service goals after those of service giants such as Federal Express. This is in contrast to the approach used by Canadian Broadcasting Corporation, which relied on industry benchmarks.

With a goal of workflow compression, Connecticut Mutual targeted an average 35% improvement in productivity for all areas and the elimination of unnecessary work processes. To accomplish this, they took the elaborate work design and analysis charts developed during the business analysis,[9] eliminated unnecessary parts, and applied technology tools to collapse the remaining process. The resulting process activity requires minimal or no hand-offs, and gives full accountability to individuals. Again we see the consistent thread of ownership as key to effective workflow.

---

[9] Refer to the discussion of the *System Schematic* for a proposed method of creating these charts.

Accomplishing workflow compression required scrutiny of information in a point-of-entry/point-of-exit analysis. This analysis put emphasis on the elimination of paper dependence and the integration of all media that users needed to do their jobs, creating a single-point-of-access environment for each worker. Connecticut Mutual also spent a substantial amount of time determining the most cost-effective and process-correct form for routing information through the organization so it is automatically delivered to the correct individual and in the right form for each task.

A change in the relationship between employees and clients is one example of how re-engineering benefited the company. Employees, who now have immediate access to on-line information about a client, now speak personally and conversationally with clients. The resulting increase in client knowledge that has resulted has opened opportunities for the development of new cross-product selling and an affinity that has increased client retention and overall satisfaction.

Company-wide Commitment Required

Connecticut Mutual's effort is also a study in collaboration with technology providers. At the outset, management recognized that for re-engineering to be successful enterprise-wide, a common architecture needed to be established. Vendors must commit to use common utilities and architecture and they were also required to sign a document that stated their intent to work with each of the other vendors involved in ONE IMAGE. Once again we see the importance of collaboration among participants, in this case extended to outside parties.

ONE IMAGE Project Goals

When Connecticut Mutual embarked on its ambitious re-engineering scheme, it set criteria, that if met, would make the project a success. They included:

- More rapid customer response. The beneficiary change process had required twenty-two human interventions. It is now done with only two and takes seconds compared to days.
- Improved operating productivity for all areas with a target of 35% average improvement. Success to date ranges from 20% to 60%.
- Collapse the business process. Connecticut Mutual is actively reviewing what it does and is applying its tools to compress the process, using business requirements as its yardstick.
- Dependence upon paper eliminated and work processes changed from a sequential to a parallel, simultaneous environment. Image processing and workflow scripting allow this.
- Integration with a command center to provide the user with all media required for his/her job—video, voice, text, and data.

Effective Strategies and Philosophies

As Connecticut Mutual handled its re-engineering, a number of strategies and philosophies have emerged which have shaped the direction of the ONE IMAGE project:

- Senior management sponsorship and participation from the start has been critical to success. The Board and top executives were briefed early and regularly along the way. Management teams signed off on the concept and were committed to openness, information flow, and training for all employees affected.
- No single vendor is responsible for the full technology package. This broke the myth and dangerous approach of an internal or external vendor monopoly. Vendors must deliver functionality, not promises. Each vendor must sign a letter of understanding up front in which it agrees to work closely with other vendors to solve Connecticut Mutual's business problems.
- Connecticut Mutual clearly stated that this enterprise-wide solution required teams that ignored hierarchy, turfdom, and rank and serial number. Cross-divisional teams of business and technical personnel were brought together, charged with specific tasks, and given the authority to complete them. Teams share resources and rotate project leaders. "Capture Business Benefit Now is the rallying cry."
- The goals of interoperability and independence have defined the solutions. For interoperability, Connecticut Mutual requires that vendors meet standards for hardware and software that cross a variety of vendors. It has required vendors to deliver product, not promises. It also declared software, hardware, and location independence. In doing so,

Connecticut Mutual has had to act as its
own systems integrator and software
shop while vendors evolve to a truly open
situation. In turn, however, this has
allowed Connecticut Mutual to work with
the best components of each vendor's
product. Through the client/server
architecture, these objectives have been
realized in a business and technical work
world that is staged, with the proper
graphical presentation for users, the
proper movement and storage of data,
and proper use of software and hardware.
Over the next three years, Connecticut
Mutual will extend its client/server
strategy and move systems off the
mainframe. The legacy systems have
become the time bottleneck that present a
difficult challenge to overcome.

• Planned communications with all employees on a continued basis reaffirms Connecticut Mutual's commitment to re-
engineering enterprise-wide, including at
field offices nationwide. Ongoing PR
includes videos, news notes, demos, and
tours. Connecticut Mutual has even wired
the cafeteria to show the command center
in action. Management reaffirms its
commitment to active, ongoing process
redesign through company meetings at
the start of each year.

The business improvements re-engineering has
delivered to date have allowed Connecticut Mutual
to downsize its staff by more than 10% and meet
expense reduction payback goals for the project.

All of this may not seem to be a crisis situation
to most readers, but therein lies the key to

understanding the true definition of *responding* to a crisis. In large part, the response is only successful when the crisis can be seen as a future event that is inevitable if certain actions are not taken. As we said earlier, if an organization waits until it is in the midst of a crisis, it is likely too late to stem the spiral of cultural, economic, and market forces that have amassed, and it is certain that a competitor has already responded. After all, an organization is only called visionary if it takes actions that other organizations have ignored. Otherwise, there would be no benchmark against which to gauge the magnitude of the organization's vision.

...the potential impact of new technology would be handicapped if Waverly did not also implement organizational changes and restructure its workflow.

Waverly Press: Workflow without Workflow

One of the most often encountered reasons for crisis re-engineering is that of being caught in the serialized, assembly line process model popularized by the post-industrial-revolution era. But not all organizations in this state associate problem symptoms, such as extended cycle times and excessive quality problems, with this well-established paradigm of the twentieth century.

When Waverly Press embarked on a master plan to revamp its composition and publication systems, it had clear goals concerning the implementation of new technology and clear reasons for undertaking a revamp; reduce cycle time by at least 40% and increase productivity by two to three times current rates. Something peculiar happened however, as Waverly began to examine related workflows and job functions. It became apparent that the potential impact of workflow, or of any new technology, would be severely handicapped if Waverly did not also implement organizational changes and restructure its workflow. As a result of upgrading its production systems, Waverly was able to make the transition from a process-oriented assembly-line structure to smaller, responsive teams that were faster, more accurate, and, above all, customer-focused. What is most interesting about Waverly's workflow implementation is that, although workflow was redesigned, no workflow-specific software was used in this implementation.

A Master Plan for Life-Cycle Document Management

Waverly Press, a leading full-service printer of scholarly, scientific, and medical journals and publications, had developed a master plan to link advanced commercial publishing technology with its existing desktop network and unique electronic copyediting system. The goal was the creation of a seamless electronic production capability. That would allow Waverly to follow a document from receipt of manuscript either in hardcopy or electronic media through final printing—a complex, multi-stage process. Waverly also wanted to tie its editing system into a content-driven database that would support digital archiving, alternative media (such as CD-ROM), on-line publishing, and demand printing.

That in itself was an ambitious undertaking, but at the same time Waverly wanted to reduce production schedules by forty to fifty percent and double (or even triple) productivity.

Modeling Uncovered Opportunity for Organizational Change

Before beginning its implementation of the workflow system, Waverly investigated and modeled various alternatives. The initial intention was to simply replace the older technology while maintaining existing support systems and workflows. This approach was uncomplicated and seemed to provide a straightforward technical transition to the new computing system. A second option was to intersperse workflow-enabled desktop computers in production areas and gradually move production staff over.

The increased computing power of the new RISC-based workstation technology and the enhanced utility of the new software represented a generational leap forward over what was in place, and would result in substantial improvements over the existing operation. Management felt simply adjusting to the complexity of the technology would be enough of a challenge initially. They thought desired organizational changes could come later, a mistake made all too often.

As Waverly examined workflows more closely, management realized that the sophistication of the new technology could eliminate many process hand-offs and would support a number of desired organizational changes. It became clear that not making these changes would actually constrain the usefulness of the new system.

Waverly finally decided upon a pilot arrangement that is among one of the most novel and successful I have seen. They established autonomous groups that were chartered to make the conversion work—technically and organizationally. Although Waverly was not aware of it at the time, they were putting into place points-of-ownership, or hybrid generalists, that would simplify tasks and collapse process times. Waverly staffed its first pilot team, or what they came to call a "cadre" (literally, a nucleus of trained personnel), with four people and a system coordinator. Once this group was fully trained in the new system, it took over two production jobs in a live mode. When these were working satisfactorily, Waverly used the first cadre to seed two new cadres. Waverly continued to subdivide cadres in this fashion until all of the production function was converted.

Old Technology Required Job Specialization

To understand Waverly's need to restructure workflows while implementing the new system, it is necessary to look at the hand-offs required with the existing production system. To compensate for the limited capabilities of the prior automated composition systems, Waverly had created specialized job functions to minimize multiple passes through the system and ensure coding was accurate. Due to limited What-You-See-Is-What-You-Get (WYSIWYG)[10] functionality, pagination was code-intensive and largely accomplished in the blind. The lack of integration of text and graphics required paste-up operators to add illustration and markup pages by hand. Waverly placed heavy emphasis on unique job functions. Production teams had developed mini departments to perform specialized roles. Specialization had gone beyond the assembly line paradigm to become an obsession.

Assembly Line Focus Limits Tasks

The system resembled a classic assembly line operation. A job packet moved progressively through specialized areas until completed. With as many as 1,600 work jackets in process through the plant at any one time, a centralized scheduling and tracking system had become the heart of production. Work was logged in and out of each successive area as it progressed through the plant. Each production area required a hand-off and a separate queue.

---

[10] This term became especially popular during the desktop publishing revolution of the late 1980s. It describes the ability to view a document on a computer monitor precisely as it will appear when printed.

Planning, encoding, and tracking requirements prevented work from immediately entering automated production. The specialization in job roles prevented personnel from seeing the impact of their efforts on the job as a whole, and by design, fostered a task focus—the proverbial plow horse with blinders. Still, despite its inefficiencies, this system maximized the technologies available within Waverly at the time.

New Technology Restructures Workflow

With workflow analysis and the application of some basic, readily available technologies, Waverly suddenly saw opportunities to reduce its dependence on role specialization. Powerful software for integrating text and graphics required less manual paste-up. More powerful job setup and style sheet capabilities required less intensive interactive coding, job planning, and encoding prior to automated production. The WYSIWYG workstation capabilities allowed a single operator to key corrections and repaginate. Reduced role specialization has meant fewer hand-offs and reduced queuing time, resulting in faster turnaround for Waverly's customers. Waverly even went so far as to relocate its laser printers, a prized and secured resource used for proofing, to the production areas where each team can produce page proofs on its own schedule without having to compete for device time in the computer room.

Through reorganization of the work process, Waverly now uses more teams with fewer members per team each team responsible for fewer publications. That has allowed each team to place more emphasis on the individual needs of a smaller group of customers. Waverly believes this change represents the most important organizational opportunity of the new system. It enabled the

The specialization in job roles prevented personnel from seeing the impact of their efforts on the job as a whole.

company to make the transition from an assembly-line structure that was inherently process-focused to smaller and more responsive teams that are customer-focused. Each team now schedules its own work and tracks the articles in its area. The new work environment is almost at a polar extreme from that which existed prior to the workflow analysis. Ultimately, the new system empowers users far beyond the artificial boundaries of the specialized assembly-line model.

Ultimately, the new system empowers users far beyond the artificial boundaries of the specialized assembly-line model.

Achieving a Customer-focused Organization

Initially, Waverly viewed the workflow installation as a way to significantly enhance its suite of composition and production technologies to improve the quality and level of service it offered customers. It openly acknowledges that the strengths of workflow software provided the vehicle to accomplish that goal.

The real story, however, is that in developing its implementation plan, Waverly recognized that the new system could be a vehicle for driving strategic change in the structure of the organization and its workflows—changes its implementors believe would have been difficult with existing systems. In many respects, this represented the most valuable opportunity provided by the new system. In fact, not making these changes while merely viewing the new technology as a replacement for the old system would have severely restricted the benefits of the installation. Instead, Waverly viewed it as an opportunity to re-engineer the work environment and improve the workflow by moving away from an outdated mode of highly specialized and discrete organizational functions. Thus, Waverly turned a potential crisis into an enormous opportunity.

A Case Study of Goal-oriented Re-engineering: Burlington Air Express' Experiences with Worldwide Workflow

Burlington Air Express (BAX) had a basic reason for looking into imaging and workflow technology—competitive advantage. Maintaining its market position required better access to information for selection of optimal delivery routes and the ability to provide customers with timely supporting information on the status of shipments.

Meeting these goals had been hampered by a paper-intensive, decentralized organizational structure. Accounts Receivable and Credit Collections faced the brunt of the productivity problems, so investigations into a solution began there. Under the sponsorship of the Chief Financial Officer and the leadership of the Corporate Controller, worldwide automation and workflow routing to critical financial and customer systems became a corporate project. The system is extensive incorporating faxing, imaging, a satellite WAN, and barcode recognition. While imaging and workflow have helped this international shipper meet its financial and customer service goals, getting to this point challenged the company's technologists in unanticipated ways.

The project began in Accounts Receivable to address specific inefficiencies that management knew could be improved if cumbersome paper processes were image-enabled. By the time the implementation was fully underway, BAX had embraced the ambitious goal of integrating imaging and workflow processing with critical corporate financial and customer systems. But doing this would also require a major cultural change within the corporation, as data entry functions moved from field offices to centralized sites.

The Workflow Learning Curve

To understand the challenge BAX faced in putting together its system, it is important to distinguish between off-the-shelf workflow, an application implemented using the vendor's software essentially as is, and customized workflow, or the adaptation of vendor software to match the unique requirements of the enterprise.

BAX describes its system as customized enterprise-wide workflow that is deeply interactive with mainframe systems.

"People will be surprised to learn how difficult it is to program workflow and make imaging work with enterprise systems," says Janet Helvey, BAX's Controller, Accounts Receivable. "This is a totally new application to most programmers and systems developers. It will challenge their skill set and you need to allow them a learning curve. You are no longer talking about data, but images and *processes* also. The concept sounds simple, but when you start plugging it together, it is a leap in many directions.[11]

"When we first attempted to integrate mainframe airway bills' data entry and image indexing—a relatively simple task—system response time was unacceptably poor. The system presented the operator with many images and when the operator selected one, it was painted on the screen, a process which impeded background indexing. The complex interaction between

[11] "Imaging and Workflow Improve Service Competitiveness and Productivity for Air Freight Company," *The Delphi Workflow Report.* March 1994, Volume 2, Issue 4, p. 2-4. Published by Delphi Consulting Group.

programs and the graphical user interface (GUI) also demanded more desktop processing power than we had expected. We found workflow is not as turnkey as people think. Especially if you are integrating it with other systems. Be especially prepared for a substantial programming effort to make the integration fit with existing databases."

Clearly Defined Goals Were Just the Beginning

BAX had a clear direction of what it wanted to do with its system. The design team had done its due diligence in months of meetings with cross-functional groups of users. Elaborate functional specifications were developed with the help of consultants and the imaging vendor. BAX had put on paper detailed designs with elaborate flow charts and organizational outlines—any one of which looked very feasible. However, it was when workflow, Windows, and the mainframe were added to the equation that many ideas became prohibitive.

"We learned to become very specific," said Helvey, "defining exactly what we wanted imaging and workflow to accomplish on each of the different systems we were using—PCs, mainframes, Windows, etc. Due to the nature of our imaging implementation, which is comprised of three systems, each with its own subsystems, decisions relating to the placement of programs and background tasking queues—where they would run and flow and how they would interact—were critical design topics. Today, we have mini-computers dedicated to monitoring background tasks and queues for over fifteen different kinds of documents. This is how we were able to consolidate data entry and still give any customer service representative anywhere on the system instant access to an airbill image and the ability to get faxed copies of the image and supporting documentation immediately."

Change Is Not without Pain

Radical organizational changes paralleled the introduction of technology at BAX. Imaging introduced local offices to a more powerful generation of desktop computer systems and scanning technology—changes that made many wary. BAX relieved local offices of data entry responsibilities with the commitment that corporate would be able to enter and index all documentation in an efficient time frame. The collection group was restructured as integrated teams of collectors, researchers, and clerical staff able to follow an airbill through to closure, demonstrating yet again the principle of ownership. Collectors now field ad hoc telephone inquiries from customers, referring to remote source documents thousands of miles away.

During the early months of the transition, however, misconceptions over the efficiency of the new system circulated in field offices. Many misinterpreted the use of imaging, and workflow, and what it would mean to them. It was a challenge to clearly explain the impact and benefits of the technology when so much was unknown until BAX went live. The design team developed new station implementation packages, held classes, and spent considerable time on the telephone training local offices about to go on-line. Each office was involved in a conference call to explain the program and establish a start date. Commitments were hard won. In the long run, it was the firm resolve of BAX corporate sponsors to meet the long-term goals of improved customer service and cost control that has kept BAX offices on track.

What Flavor of Workflow?

The way Burlington Air Express approaches applications today reflects experience and maturity with workflow. The first subsystems evolved through many prototypes and interactions with users. Developers built one function at a time according to written specifications. After having a number of early efforts rejected by users in prototype, the developers found it best to use the vendor's workflow as a springboard.

"We used the vendor's software as the basic structure for the workflow, a structure which we ultimately changed as we configured our system to meet our needs," explained Helvey. "Now that users, management, and application developers have a better idea of the capabilities and limits of workflow, the first cut on new functions requires only minimal revisions. Our experience is also allowing us to build several functions at the same time. Our design team still meets *weekly* with users and vendors to review and discuss applications under development—users view work-in-process in a prototype mode and critique it. Today we program a little, review a little. The nature of workflow—in which many functions feed other functions—requires us to do as many things as we can simultaneously. Experience has made this possible."

The nature of workflow—in which many functions feed other functions— requires us to do as many things as we can simultaneously.

"It has taken us about a year in time, including the programming learning curve, to get to where we are today," reflects Helvey. "The outcome is worth it. The imaging and workflow system has significantly shortened the time for service representatives in any office on a package's route to locate important delivery documentation. We are more efficient in handling the paper and documentation needed to invoice. Customers are experiencing an 80% improvement in response time and are now able to get the supporting documentation they require. Cash flows have improved and we have achieved improvements in productivity and a significant reduction of data entry costs from the re-engineering."

Quite possibly the most significant benefit is that BAX has been able to process huge increases in shipping volume with a constant head count. And in the spirit of life-cycle re-engineering, development has become an ongoing process as the organization and its processes evolve.

The most significant benefit is that BAX has been able to process huge increases in shipping volume with a constant head count.

Commentary

Integrating workflow with enterprise systems takes expertise, detailed planning and a solid corporate commitment. As we have seen repeatedly, it also takes teamwork between management, information systems (IS), and the users. BAX was ambitious and aggressive in taking on a project of this magnitude as its first workflow effort. It succeeded because things were done right from the beginning.

First, BAX took the time to develop an appreciation of its current processes and situation. It is virtually impossible to create a successful workflow implementation without first understanding existing processes.

Second, BAX developed an appreciation for the technology and recognized that workflow is different. It requires new skills and new analytical approaches and thought processes. Also, there is a learning curve.

Third, users were involved early in the process. Users must learn and use the new system and procedures, so success would be impossible without their cooperation. Even so, it took a lot of planning, hard work, and training to get the proper level of user participation and buy-in.

The final success factor was dedicated, high-level sponsorship with clearly defined goals that were communicated well and often. Without the commitment and leadership from management, the new infrastructure, organizational and cultural changes, and, the belief in the necessity of change simply would not have happened. BAX could have ended up with an expensive, but ineffective, automation exercise.

Which Approach Is Right for You?

Making a Decision

Which approach should you use? That may have already been decided for you by the culture of the enterprise or the state of your information systems. If you are concerned about the long-term impact of crisis re-engineering, find a corporate sponsor who is willing to support a goal-oriented approach and help break the crisis cycle.

Whatever else you do, do not embark on a perpetual journey in search of sponsorship.

Whatever else you do, do not embark on a perpetual journey in search of sponsorship. Far too often, organizations chase their tails in an endless series of good ideas that do not have the backing of sponsor. Just about everyone would like to have the CEO or chairman behind a particular project, but, that is rarely the case.

If you can't find a sponsor, maybe you are looking in all the wrong places. If there is a departmental manager, a divisional VP, a person in or any other position willing to stand behind a smaller effort, take advantage of it. It is likely to be the only way to develop higher levels of sponsorship.

Over the long-term, the sponsor must establish an ongoing review of the business process. A workflow system that includes management reporting tools will help identify problems and ongoing areas of improvement in the process model.

Critical to every re-engineering effort's success is the ability to overcome the cultural impediments that you are bound to encounter. Although there are creative ways to deal with these, they are generation to a large degree. They can only be extricated over time through investment in education, some attrition, a substansive change in the work environment and work tools, and a cohesive commitment to change from management.

Don't expect to re-engineer or automate the workflow of an entire enterprise overnight. The battle-worn will tell you of the upheaval an organization goes through to change work patterns and business practices. An incremental approach provides an education and a demonstration of the benefits to be realized. An approach that includes measurement of the existing and new process models provides the greatest impetus for change and long-term life-cycle re-engineering. Workflow makes that possible.

Ultimately, the ability to uncover new potential from business opportunity, not the application of technology to current problems, makes re-engineering worthwhile. This is a common theme in virtually every successful workflow implementation. That potential will seldom be realized without a tool such as workflow—which is able to locate and to measure the extent of problem business processes, at each point along the enterprise life-cycle.

The battle-worn will tell you of the upheaval an organization goes through to change work patterns and business practices.

Designing Workflow Applications: A New Perspective

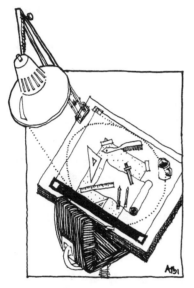

Implementing effective workflow is as much a matter of adopting a new perspective of your organization as it is understanding the technology.

Implementing effective workflow is as much a matter of adopting a new perspective of your organization as it is understanding the technology. With an understanding of both, you are then ready to put in place a revolutionary new tool for process automation. This chapter discusses a set of methods for the analysis of an organization's workflow requirements. The benefit of these methods is to help you gain a new perspective of the organization. Although that sounds simple enough, it requires a level of objectivity that is not likely to be commonplace in your organization.

In that light, the considerations, methodologies, and analytical techniques that are presented
in this text have been developed to provide a
perspective that will help you to address the specific
challenges of implementing a workflow system.
They have been developed in the field, worked and
reworked in the context of many workflow analysis
and implementation efforts. Projects used to develop
these methodologies and techniques range in scope
from turnkey projects of only a few hundred
thousand dollars to complex workflow projects
with investments in excess of five million dollars
over a three-year period. They involved industries
as diverse as railroads and aerospace. In all of these
cases the fundamental methods you are about to
explore have been applied across this range of
diversity with amazing consistency and repeated
success. The key ingredient of success has been the
objectivity that the process brings to an organization
that may have been recycling its own assumptions
for far too long.

Tools for Designing a Workflow Application

The tools described here are intended to establish
objectivity; therefore it is essential that you use
them as part of a team exercise. In developing these
tools we have attempted to provide a framework
for not only collecting information about a business
process, but also to present that information to a
team in a clear and understandable format.

A good example is the first tool discussed, the
System Schematic. It is a relatively straightforward
and up-to-date graphical representation of your
hardware, software, and networking infrastructure.
This may be one of the most critical elements to

have in place prior to discussing an organization's current or planned workflow, but few organizations have gone to the trouble of engendering their information systems, or staff, with an automated ability to produce such a basic rendering of their systems architecture.

Imagine trying to build a highway without first having an intimate understanding of the topography and geology of the terrain. It would be an exercise in futility. Highways are not straight for two reasons: it is more cost-effective to follow the terrain than it is to ignore it, and straight lines ignore the human elements of driving—namely, that you would fall asleep at the wheel if something did not require your ongoing attention.

The same can be said of information systems. They must work within the legacy and they must acknowledge the way users navigate through information. Yet these basic tenets are easily ignored. Many disciples of re-engineering believe, unfortunately, that defining an end-state, and committing the organization to it, is sufficient to traverse all obstacles. Given unlimited time and funding, that may be true. The reality is that time and money are finite. It is easy to pick points A and B—your current-state and your end-state—the difficult part is defining the route by which you will get from one to the other within these finite constraints.

Why then do I always get a resounding cry of reluctance when I step up to a client's drawing board and begin the process of creating a System Schematic?[1] Because I have yet to work with people

---

[1] A metaphor. In an actual engagement I would never recommend the use of a white board, flip chart, blackboard, or other such primitive means for creating a System Schematic. There are a variety of software tools available in the market by which to do this. These offer infinitely greater flexibility and longevity in creating a living Systems Schematic.

who believe that they do not have a firm grasp of what makes up their organization's infrastructure. And every one of them is wrong! Systems change rapidly, and with desktop computing becoming a commodity, there may be no central control of hardware, software, and communications systems. Even if you have a Systems Schematic in place, ask yourself the following: How old is it? When was it last updated? Have end users seen it and commented, or is it a product of IS staff only?

Other methods, such as Stair Step—a tactical approach to change management—may appear to be entirely contrary to the prevailing popular belief in re-engineering. Time-based Analysis, which shifts the focus of process redesign from workers to process flows between workers, may seem non-intuitive at first.

As you work through each of these, however, keep in mind the value of objectivity and methodologies; simply put, they are enablers for collaborative thinking. They validate your intuition, in some cases, and challenge it in others. Although these tools set the tone and framework for collaboration none of them, or any other methods, can replace sponsorship, end user buy-in, and good judgment. The bright side is that no matter what the outcome of your initial effort at workflow, establishing the theme of an agreed-upon framework for use by an objective team will carry you through the many unanticipated obstacles that you are bound to face.

Last, all of these methods can be applied in a short time frame, and always in parallel with a standard analysis that you would undertake in the prudent development of any new system.

Setting the Course

The first step in implementing workflow is to develop an understanding of the obstacles that implementors face. These center on four primary areas; infrastructure, application selection, process redesign, and organizational factors. Each area accounts for a significant component of the workflow analysis.

Infrastructure integration is discussed in the section on the System Schematic. Process redesign is addressed in the section on Time-based Analysis.

We have spent considerable time discussing an array of organizational and human issues that must be considered; now we will see how they play a role in both Stair Step and Time-based Analysis. Here is a quick overview of each of the methods we are going to cover.

The first phase, defining the existing business process and technology infrastructure, is accomplished through the creation of a System Schematic. The System Schematic provides a comprehensive definition of the current components that make up the tasks and information systems used in the organization prior to the application of workflow. It uses a graphic representation of the present system's technology infrastructure and the flow of information within the present structure. This provides a clear framework for the discussion and examination of problems and alternative solutions.

The second phase, selecting the initial application, uses an enterprise modeling technique called Stair Step. The Stair Step model provides a context for identifying the most appropriate sequence of precisely defined workgroup applications, the first of which will be the pilot application.

One of the more interesting aspect of Stair Step is its reliance on a new method of organizational grouping, the *Workcell*. Workcells consist of interfunctional teams that are organized spontaneously for a specific process. These teams may be established sets of individuals or skill sets. This reflects the nature of what we earlier referred to as a perpetual organization, constantly changing and ignoring the rigid boundaries of traditional hierarchical or networked organizations.

■■■■■■■■

Stair Step provides a foundation of experience, which is the most difficult and important ingredient of any workflow application.

Stair Step also stresses an incremental approach to developing enterprise workflow systems. The method takes into consideration the complexity of large workflow applications and the dynamic nature of most enterprises.

The benefit of Stair Step, especially in large organizations, is its ability to regulate expectations and demonstrate results quickly, allowing the technology to gain favor among both users and management. Finally, Stair Step provides a foundation of experience on which to build enterprise applications and to strengthen communications among workgroups, which is the most difficult and important ingredient of any workflow application.

■■■■■■■■

Time-based Analysis focuses on the portion of the business cycle between people.

The third phase identifies the areas of business cycle weakness and inefficiency. It uses the results of the System Schematic along with a process for establishing the timing of each business cycle to target areas of potential productivity improve-ments. This methodology is called Time-based Analysis. It focuses on the element of task versus transfer time, which is absolutely essential to a proper workflow analysis and the most often ignored aspect of many approaches that use traditional data-flow modeling techniques. By using Time-based Analysis, it is possible to achieve quantum gains in productivity without massive re-engineering, as we saw with the case study of Waverly Press.

Empowerment:
The Reality

We hear far too much about the benefits of empowering workers through technologies such as workflow and not enough about the pitfalls of empowerment. Many first-time implementors of workflow are realizing that the very people they need most to empower are the ones least capable of empowerment and most resistant to change. The classic example is that of the workflow-enabled mailroom. Correspondence is routinely scanned and routed to recipients, rather than routing the actual paper mail in its envelope. The problem arises in that an additional level of skill is required to make determinations as to what should be routed electronically and to whom it should be routed. The net effect is an additional burden on the worker to understand the new technology and an initial decrease in productivity.

The result? Empowerment and change happen gradually, not as the result of total re-engineering. But without an organizational mandate to change, what's the solution? Many organizations have figured out a way to do just that, without the risk of disruption and resistance that often accompanies re-engineering. Here's a glimpse at what they are doing and why.

The fact is that change is a buyer's, not a seller's market.

First, and foremost, these organizations have determined that change is a buyer's market not a seller's market. Enormous change cannot be inflicted on an organization without enormous frustration and turmoil on workers. Incremental change provides a template, education, and track record for long-term success. This seems to be leading in the direction of a two wave workflow market for vendors of workflow technology. The first wave is characterized by sales to small workgroups of twenty-five to fifty people. Although most first-time installations still fall in that range, workflow is expanding in many enterprises to applications with hundreds or thousands of workflow-enabled seats. The staged nature of this expansion, however, is an interesting one to analyze.

Workflow does not require drastic change.

Rather than embarking on an initial workflow application that is larger in scope with several hundred immediate workflow seats, evaluators are looking at long-term workflow implementations that will result in 1,000 plus workflow seats over a two-to-three-year time frame. This new market model for workflow is significant in that it spells out a longer-term strategy for workflow applications in large organizations. In addition, this model allows aligned incremental change by enabling smaller workgroups, coinciding with a long-term direction of strategic change. The result over time is a single solution rather than myriad independent solutions chosen by each separate workgroup.

Also of importance is the acknowledgment that workflow is increasingly becoming a technology that is implemented in a groundswell mode among small groups of individuals. Despite the dramatic benefits of complete re-engineering efforts, the reality remains that most organizations, (in fact almost all that we have surveyed thus far) talk of a kinder, gentler approach to change management and workflow implementation. That may not be what the purists want to hear but it is what the market is saying. This is also one reason why so many workflow vendors have been bucking the trend towards creating a synonymous relationship between workflow and re-engineering.

This will have profound implications upon the workflow applications market. The market may become segmented into two clear categories of products, and perhaps vendors as well: enterprise workflow products that rely on a long-term re-engineering premise and ad hoc workgroup workflow products that do not require the enterprise commitment or sponsorship of re-engineering. Indications are that the market for workflow technology will grow, without prejudice to either camp, and that many vendors will have to bridge both of these market segments, either through their own offerings or underlying standards, in order to provide integrated enterprise workflow solutions.

The System Schematic

As with most automated systems, the first step in designing a workflow solution begins with an understanding of the work environment, the work processes, and the users' needs and requirements. At this point, we assume that the reader has a basic understanding of workflow systems, which includes a repertoire of the available products, methodologies, applications, benefits, and limitations.

The System Schematic is the foundation of a well-designed workflow application. It will be used throughout the entire analysis and design process to assist in understanding how information flows through the organization, how it is processed and accessed, and how the proposed workflow environment will be supported by existing or planned hardware, software, and communications infrastructure.

The primary objective of the System Schematic is the development of a common understanding of the organization's existing technology infrastructure among users, evaluators, and implementors. The term infrastructure denotes the full range of all hardware, communications, and software that make up an organization's information systems, including paper and file cabinets.

The secondary objective is the development of a framework that identifies the major areas of concern and potential difficulty in the workflow implementation. These may be found in the lack of sufficient communications between distributed nodes, or the inability to support the variety of end user platforms in place.

The System Schematic is, at first, an orientation tool that can set general expectations about the scope of the workflow project and the degree to which existing information systems will support desired results. In its final rendering, the System Schematic acts as the foundation for establishing Request for Proposal (RFP) requirements for workflow and the development of a project plan.

One of the most interesting aspects of the System Schematic is how few organizations already have one in place. For the last five years I have been asking audiences at my seminars the question, "Do you have a complete and up-to-date graphic schematic of your organization's hardware, software, and IS infrastructure readily available, so that you could produce it within one hour if I were to walk into your office and request one?" Amazing as it may seem, less than 2% of the attendees have answered yes. The remainder simply do not have such a road map of their organization.

Without this, you start with two distinct disadvantages. First, your assumptions about the existent workflow will be based upon heresy and fragmented perspectives, rather then the actual infrastructure in place for facilitating the workflow. Second, because workflow can have an enormous impact on access to existing applications and will be limited in its process scope by the availability of existing systems and platforms, you will need up-to-date detail of your information systems infrastructure in order to assess the availability of workflow technology alternatives or required modification to the infrastructure.

It is not unusual, however, to encounter resistance to the process of documenting the existing infrastructure. The incorrect perception is that by doing this you will be paving the cow paths. The fact is that a workflow system either, supports the variety of infrastructure components already in place or, it requires a new infrastructure. Without a

One of the most interesting aspects of the System Schematic is how few organizations already have one in place.

System Schematic in place, assumptions will be made about the infrastructure and the business process that fail to manifest existing inefficiencies. In addition, there is no denying that much of the legacy in applications and data that you have in your organizations resides in the existing infrastructure. Ignoring this will undermine the process redesign. The single largest problem faced by organizations considering process automation is the support or conversion of these legacy applications.

Interestingly, there is a clear distinction between technology evaluation and investment inside and outside the United States. Outside of the U.S., there is a clear unwillingness to invest in technologies that abandon legacy applications. In some ways that may be considered an anchor to the past. It certainly clashes with some proponents of re-engineering who advocate radical extrication of old processes. At the same time, the apparent disregard that abounds in the United States when it comes to recreating solutions is almost reckless. The two extremes have been described by some as a difference in technology investment strategy.

It is probably safe to say that neither extreme represents sound change management. Sound change management weighs not only the cost and benefits of rebuilding, but should also weigh the costs and benefits of preserving the existing process and information assets.

Instead of razing old structures each time we build new ones, the focus should be on the value of preserving these organizational assets, whether in the form of process knowledge or information. For example, a hospital's patient records system may consist of a variety of paper and electronic documents. Although a day-forward workflow management system could be put in place for all

Sound change management weighs not only the costs and benefits of rebuilding, but should also weigh the costs and benefits of preserving the existing process and information assets.

new patients (and it is certain that it would streamline admittance and provide higher quality health care) the legacy of patient records will require a decision to either convert these records to the new system or integrate the old system with the new process. Which is better? Without knowing intimately the nature of the old system, the frequency of re-admitting patients having existing records, and the costs and benefits of each approach it is impossible to tell.

That's precisely the point of the System Schematic. Making an assumption that there is a *right* answer, before evaluating the existing infrastructure and process simply because you know where you want to go, is paramount to skydiving without first checking wind conditions, altitude, and equipment. You'll go in the right direction and arrive at your destination one way or the other, but not in the same condition.

The System Schematic is constructed from information gathered by interviewing key individuals participating in the evaluation of the workflow system and users of existing information systems to be integrated into the workflow environment. A peculiarity of the approach used to create a System Schematic is that the initial interviews are not intended to determine the needs and requirements of users for the new workflow system. The first step is to simply develop a better understanding of the existing technology infrastructure.

The initial System Schematic, as you will see, need not be elaborate; it should simply serve as a framework for validation of assumptions about systems and process infrastructure. There is no need in the early stages of the System Schematic to provide extensive detail about the processing of the information and every role involved in its

processing. This will be undertaken later in the
evolution of the System Schematic when it is used
as the baseline for Time-based Analysis.

Phase I: Defining the Project Scope and the Organization

Developing the System Schematic is a three-phased
process that begins with the definition of a project
scope and an assessment of the existing information
systems infrastructure. The project scope can be as
broad as solving an entire organization's document
management needs or as narrow as the development of an application to automate a single
workgroup.

The best way to define the term *organization*, in
the context of the System Schematic, is to ask the
question, "What portion of the organization's IS
architecture falls under the highest-level corporate
sponsor for this workflow implementation?"

This definition of *organization* provides a practical scope that can be managed and controlled by
the presence of a singular sponsor. It also identifies
the success criteria and realistic expectations—both
of which are absolutely essential to your success.

As we have already said, the sponsorship should
be such that this individual is not only committed
to the workflow application but also has the budget
for workflow or, at least, the ability to justify
workflow to a higher budget authority. Without
this level of sponsorship, and without a clearly
defined scope, it will be next to impossible to gather
sufficient momentum or build a substantive case
for workflow. The reason is that so much of
workflow's benefit is in the productivity gained
through the development of new work models.
This is not easily justified prior to implementation.
You must therefore gain sponsorship and

implement within an organizational scope that will provide adequate support to demonstrate the value of workflow. This piece of the organization will be the principal target for workflow implementation. It will not, however, become an exclusive application that ignores the remainder of the enterprise and users outside of the initial project scope.

We will attempt to balance the requirements of the pilot application with the overall requirements of the enterprise through the use of the Stair Step methodology. However, we should be clear that our main objective is to meet the success criteria for our organization and create a successful benchmark for future implementations by other organizations. Do not step out of this scope and compromise the chances for a quick measurable success by constantly deferring to the perceived requirements of peripheral applications and organizations.

If that sounds like a fragmented approach, it is—if your sponsor's organization is not synonymous with the entire enterprise, which is most often the case. Think of it in the following way. Each part of your enterprise looks to your initial application as a test bed for their own applications. The success of this initial foray into workflow is the greatest testimonial to extending its use to their organizations. Its failure will obviously serve to do just the opposite; and that definitely would create fragmentation as alternative solutions are found. On the other hand, your success may serve to solicit a higher-level sponsor who will spearhead an enterprise-wise or cross-organizational effort. Again, failure would hardly serve that end. The key then is to identify the organization by identifying the sponsor. This is why it is so important to set expectations within the scope of your sponsor's organization.

Gain sponsorship and implement within an organizational scope that will provide adequate support to demonstrate the value of workflow.

The success of this initial foray into workflow is the greatest testimonial to extending its use to their organizations.

Phase II: Creating the System Schematic

In all cases, the process for creating the schematic is the same. The first step is to define a system architecture for the hardware, networks, and applications that comprise the available environment for the new workflow system. This will serve as the general framework for the System Schematic. A number of targeted interviews will be conducted at this point with IS, users, sponsors, and any individuals considered representative of the organizational processes that have been targeted. If multiple pieces of an organization are being considered as candidates, the System Schematic must span each of these. The level of detail you are trying to achieve in describing the system schematic can best be described as a high-altitude view of the infrastructure and process.

At this time it is also worthwhile to begin gathering routing information. Routing identifies the types of information sources, the way in which this information is used, and the path the information takes through the process. The initial routing definition is concerned only with the existing documents. It does not consider any potential changes in the way the information is handled. In most cases this will require expanding the System Schematic to include non-computer-based tasks and manual or semi-manual document management activities. Once a basic System Schematic has been completed, including a routing definition, you can proceed to the next step.

Note that at this stage the System Schematic is far from complete and, in fact, is almost always going to be a misleading representation of the actual and final version. You should resist the temptation to finalize the System Schematic at this stage. First-time interviews can never yield all of the information

The level of detail you are trying to achieve in describing the system schematic can best be described as a high-altitude view of the infrastructure and process.

needed to provide a conclusive description. Don't be concerned about this just yet. If you do attempt to finalize the System Schematic, you will make numerous assumptions and subjective analyses based on incomplete information. This is one of the biggest mistakes made by both novice and experienced business systems analysts; attempting to find "the answers" through extensive interviewing. It just won't happen. What will happen is an enormous amount of time passes, you incur high costs for consultants and their reports, and users tire of interviewing with no results.

We will take a much faster route using the System Schematic using several iterations of interviewing in several settings. You will perform at least two additional iterations of the interview process to complete the System Schematic. Later in the process during Phase III, we will return to the System Schematic in order to correct the many flaws and incorrect assumptions that have been made so far.

Interviewing

Interviewing users directly will certainly offer the best insight into their work environment. Without adequate preparation, however, the process will be frustrating for both the users and the interviewer. Without full disclosure from the users, you run the risk of having to repeat the interview process several times before developing a complete System Schematic. It is also likely that users will, at best, tire of the repetition, or at worst they will lose confidence in your ability to understand the process. This last point is a risk of any systems analysis. However, it is particularly damaging to workflow analysis due to the level of user buy-in required to successfully implement a system.

One of the qualities of a good interviewer is the lack of an ego. (Actually, the ego is probably there but it is kept well hidden.) The interviewer should avoid the temptation to stand up and say, "Eureka! I have found the solution!" First, that sort of an approach will certainly sour users; after all its their process and they have been trying to understand it for years. Who are you to suddenly open their eyes to it? Secondly, its their buy-in you are looking for. Convincing yourself is easy, but how are you now going to convince them. (Now back to my first point, and you'll stay in that loop for good or until someone comes along to put you out of your misery.)

The interviewer should avoid the temptation to stand up and say, "Eureka! I have found the solution!"

Users must not only be a significant source of information for the analysis, but they must also believe that they are a strong contributing influence. Without this sense of teamwork, you run the risk of creating a workflow application that is often circumvented, undermined, or just plain short-sighted.

The goal of the designer/interviewer is to *identify* weak areas, or potential points for collapsing current business processes, and then present these to the organization as a statement of fact, not opinion. Again, recall that we said the value of this process is achieving objectivity and collaboration. It is not uncommon to ask the users/interviewees where they perceive problems and where they would like to see change. The answers provided, however, may only identify symptoms of these problems. Your job as the interviewer is to probe beneath the surface by watching for common threads to emerge in the System Schematic and then use the System Schematic to represent these problems in an objective fashion.

These problems may speak to a particular aspect of the existing system that is inhibiting access to information, such as the lack of adequate network bandwidth for transferring documents or the inability to track document status when a document is transmitted electronically. Users in many cases will not directly identify this problem because of their acceptance of the existing system's parameters. Interestingly, they will usually acknowledge it when presented with such a finding, but confess to not believing it was something that could have been changed.

The interview process at this stage is still dealing at a high level with personnel who are not necessarily involved with the day-to-day intricacies of IS. Many users feel the need to respond to questions about the development of new computer-based systems with technical observations about the current system's deficiencies—not unlike the tendency most people have to use a limited medical vocabulary when talking to their physicians. This limits the value of the user's observations by confining their perceptions to what they believe is possible. In the context of traditional systems analysis this is actually an advantage, because users and developers can establish a common rapport and expectation set. In the context of a workflow analysis, however, user presumptions based upon existing knowledge of what systems can or cannot do is restrictive and counterproductive because it ignores the benefit of changing work models in new ways that have no counterpart in existing systems.

One technique that is helpful with users who are reluctant to offer an opinion based upon their lack of technical know-how, is the *blue sky method*. When you "Blue sky" with a user, ask them to set aside their preconceptions about information systems and describe an ideal scenario; the sky is the limit! For example, when asked to describe how to improve on a customer support system, one user replied:

"It would be nice if I could just find out which of my co-workers received the same question. Sometimes you'll get one customer calling several different reps because they don't like the answer they got during the last call. I usually find out about the question later in the day, but by then it's too late."

In this case, the problem was impossible to solve with the existing system since calls were not logged on-line during the day but at night after they had been coded and keyed for retrieval. That same user, when asked for a solution to the problem, responded:

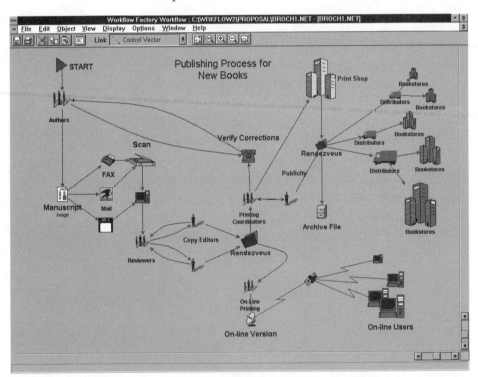

Example System Schematic produced with an automated diagramming tool.

"I would just write down all of the important calls for the day on a bulletin board in the customer support Bullpen [all of the customer support reps at this location shared a common area called the Bullpen] so that you could just glance over to it and then go to the rep who took the call and talk about the problem with them immediately."

It would never have occurred to the user that the calls could also be available through a workflow system as they were received. This type of information can be invaluable to an educated interviewer, but it must first be offered up by the users. Blue skying may be one way to encourage this. Be careful not to recommend a specific technology solution at this stage. That will serve to set expectations that you do not yet know are reasonable or appropriate and will be lured into designing systems based upon the requirements of only one or a few users with whom you have had a good rapport.

...no amount of interviewing will provide a complete picture of the organization's full information infrastructure.

Throughout this process you have been modifying the System Schematic on an ongoing basis with each successive interview. In this way the schematic has slowly evolved to reflect the organization's actual infrastructure and the transfer of information through the organization.

As we said, certain trouble spots have now begun to emerge as prime locations for automation and streamlining. Again, you should have avoided the temptation to create a finalized representation of the schematic at this time. You now have a System Schematic that you believe represents the collective view of the organization's existing infrastructure and workflow.[1] So, what do you do with it next? Here comes the fun part.

Phase III: Finalizing the System Schematic and Identifying the New Infrastructure

The final phase of constructing the System Schematic is the most valuable. Recall that earlier we said the System Schematic is always incomplete when it is first created. In all of the years I have been working with end users to define information systems, only one technique, the one we are about to describe, has worked in creating a precise definition of current infrastructure and workflow. The process is a simple one—it almost seems too simple to have so much value—yet it is a difficult process to execute because it involves a roomful of human dynamics and requires an objective facilitator who can challenge long-standing assumptions, has confidence in the process, and a bit of experience managing group sessions.

---

[1] There are a variety of tools on the market with which to develop electronic System Schematic renderings. On-line tools are recommended since they can be used to dynamically alter and reconfigure the graphical representation easily. In addition, a tool that uses vivid graphic symbols of people, computers, and other process components is preferred over a standard block diagramming tool or one with a limited set of icons. Users will respond best to a graphical tool that is immediately recognizable as their process. That will be especially valuable during phase three of the System Schematic when users are involved in validating it as a group exercise. The System Schematic shown on page 120 was developed with Delphi's Workflow Factory.

What is this simple process? A group session during which you will review the System Schematic created with no more than twenty-five[2] of the individuals involved in the interview process. The basic premise of the review is that no amount of interviewing will provide a complete picture of the organization's full information infrastructure. There are likely to be numerous situations where informal methods have evolved for the transfer and sharing of information that are not adequately reflected until the schematic is reviewed by the entire group.

The review of the System Schematic is a dynamic and important part of the overall methodology. In fact, the principle benefit of the System Schematic, as was stated at the beginning of this chapter, is the development of a common understanding of the organization's existing technology infrastructure among users, evaluators, and implementors. It is impossible to achieve this without thoroughly critiquing the perceived infrastructure, and the perspective, real and imagined, of the existing infrastructure and workflow.

The review must be conducted as a group session consisting of members from each functional group involved with the proposed workflow application and many of the individuals interviewed during Phases II. It is inevitable that many assumptions about the way the existing system worked, or was supposed to work, will be invalid. The review will serve to identify the reality of the organization's infrastructure. The review process also double checks the findings of the interviewer/designer, and catches any false assumptions that may have been made along the way.

It is inevitable that many assumptions about the way the existing system worked, or was supposed to work, will be invalid.

---

[2] The reason for the limit of 25 has to do with the nearly impossible task of managing a group discussion with any more than 25 people.

Be prepared to constantly challenge assumptions that have been made.

Be prepared to constantly challenge assumptions that have been made. When an assumption is found to be incorrect, change the System Schematic on the fly to reflect the correct view of the infrastructure and workflow. (This is why an on-line tool is probably best.) Most important, moderate the many discussions that will ensue and bring wandering discussions back to the focal point of the System Schematic. By doing this, you can almost remove your own bias, and that of anyone else from the process, by presenting the entire group with a single representation that must be agreed upon.

The result of the review is a completed System Schematic that precisely depicts the computing environment and the obstacles impeding the flow of information within that environment. At this point you can begin to assess the viability of the existing infrastructure with confidence in your starting point.

As you proceed with the rest of your evaluation, specifically using the Stair Step and Time-based Analysis methods, the System Schematic will continue to change so that it reflects issues such as storage, networks, servers, desktop upgrades, user environments, and administration. By the time you are ready to implement a workflow solution, the System Schematic will have gone through many iterations and incremental changes. But even after you have chosen a workflow solution, it is a good idea to continue updating your System Schematic to reflect the changing organization. Schedule regular reviews of it by groups of users and IS staff. Continuing this process of reviewing the Schematic will provide an ongoing system of process validation; it will establish a means of stemming process modifications that may be short-sighted and inefficient; it will offer users a broader perspective than just their own tasks; and it will foster an attitude of empowerment, wherein all users can question a process and participate in actively changing it. With this done, the System Schematic will become a cornerstone of your *Perpetual Organization*—always ready to change and reconfigure itself for the demands at hand.

The final step of phase III of the System Schematic is actually the first step in the Stair Step method; defining the pilot application. Although you have already identified the sponsorship and the project scope at the outset in order to develop the System Schematic, you have probably encountered numerous applications and workflow cycles within this project.

Now you must choose among these for your first implementation. If, on the other hand, you were fortunate enough to remain with only one application, Stair Step is not necessary. However, you should be careful not to railroad your thinking with a predisposition for one application. It is often the case that, as you work through the System Schematic, the application you first thought was an ideal workflow pilot has become unwieldy or simply did not turn out to have the payback you expected. (The payback may not be apparent until you complete the Time-based Analysis exercise.)

The pilot you choose should be a concrete, well-defined, and confined application. The System Schematic alone, however, does not provide sufficient information to select the best candidate application. It simply lays out the components of the applications under the sponsor. A further exploration of the business cycles and the potential applications must be conducted before defining the pilot or selecting from the candidate pilot applications. This is the subject of the following section, the Stair Step Method.

The Stair Step Method

As in the development of the System Schematic, the scope of the workflow application may span a large portion of the enterprise's information systems. This leads to a classic IS problem—where to begin. This question cuts far deeper than the obvious. Asking where to begin and consternating over this decision implies, in part, that the beginning point differs substantially from all other points of application. Although, at any point in time, a priority can be assigned to a variety of applications, indicating which has the greatest payback if automated—these priorities are constantly changing. That would not be a problem, where it not that the priorities often change before the applications are implemented. This means that larger units of change are accompanied by a greater risk of obsolescence.

Organizations that have pursued such large scale change find themselves most in need of periodic large-scale re-engineering.

Such situations are ironic in that organizations that have pursued such large scale change find themselves most in need of periodic large-scale re-engineering. The predicament creates an endless struggle to create a best-of-breed solution, which becomes inevitably obsolete shortly after, if not before, it is complete.

The solution is to stop this ceaseless quest for the right solution, and instead focus on the process of constant solution refinement. What sounds like a short-sighted approach is actually just the antithesis.

The Japanese have used this approach for some time. Taking a longer view, they believe the best way to achieve results is to perfect the process upon which results are based. They have also approached the process of improvement in a very conservative way, becoming great believers in taking small, incremental steps and pursuing the goal relentlessly over time. The Japanese have a popular word for this approach: *kaizen.*

This incremental approach is only possible to undertake with a long-term goal in mind. Workflow makes this possible by not only providing a binding element for constantly monitoring a process and modifying it, but also for integrating the many processes that make up the increments of change.

The issue now becomes one of identifying the incremental steps to change. This is where Stair Step is used. Systems analysis and business analysis has always employed two distinct approaches to answer this question; enterprise modeling and prototyping. These two basic approaches take many forms. For example, top-down design and programming is a version of enterprise modeling in which problems are solved from a broad framework to a more detailed one. Bottom-up design and programming uses the alternative prototype method, where individual solutions are tested and implemented before moving onto a broader framework. There are problems with both of these approaches if one is used to the complete exclusion of the other.

A step is a process or application that is defined well enough so that it can be implemented quickly and easily.

In the case of workflow applications, enterprise modeling not only requires a substantial analysis due to its complexity, but also compromises the highly individualized nature of each workgroup's workflow needs. In addition, it is difficult because the underlying office automation technologies, which need to be integrated with the workflow solution, are changing rapidly. Therefore, any enterprise system designed to be implemented all at once will be outdated by the time it is defined.

On the other hand, prototyping can result in a number of applications, each for a very specific set of needs, but with little consideration for communication and sharing of documents from one application to another. That is, in fact, how we have created the problem of fragmented and discrete solutions for current information systems.

The Stair Step Method combines these approaches, stressing a long-term commitment to the enterprise but delivering short-term results within a limited and manageable framework. The method takes a top-down approach to system design and a bottom-up approach to implementation. The Stair Step Method begins with a broad view of the problem and the solution. However, solutions are implemented one step at a time. These steps are defined through a rather simple statistical process that measures the relative difficulty and benefit of an array of potential workflow implementations. The applications are categorized by both *workcells* and *applications*.

A *step* is a process, within a sponsor's organization (using the definition of organization explained earlier), that is defined well enough so that it can be implemented quickly and easily. The principle benefit of a step implementation is that it provides quick increments of change. These increments are not discrete applications, but rather orchestrated implementations across an entire organizations workflow. As with the System Schematic, the Stair Step will be a living model of the changing organization. It will be updated constantly to reflect the true nature of the interaction between workgroups and applications within the organization.

There is no set definition as to how large or small a step is. In each situation a step can be defined differently. In one case it could be a department-wide system and in another it could be one workgroup within a departmental system. Although our definition of an organization will not change, we will use the workcell structure as a new element of the organization that is a much more accurate and current view of work than that of a department or other such traditional organizational structure. The *workcell* will be the principle building block of a stair-step model and of the workflow definition.

Defining the Workcells

After the organizational view of the infrastructure and the workflow are determined, using the System Schematic, the individual workcells that make up the processes are defined. A good example of a workcell is the formation of a team that consists of representatives from accounting, sales, and marketing. This workcell approves advertising campaigns, so we will call them "Approvers." If we were to list the possible applications for an organization that is involved with creating advertising programs it would include the *application* "New Ad Program" and the *workcell* called "Approvers." It would also include numerous other workcells, some of which may correspond with traditional departmental groupings. The point of a workcell approach is not to obliterate any and all existing structures but rather to allow for the representation of a workflow in its most natural form. In reality, most organizations do not follow the boundaries of departments and formal workgroups as often as managers would like to believe. As a result processes take shape over time that run contrary to the accepted myth of an organizational hierarchy. Many of the incorrect assumptions about process steps are perpetuated because of this. Your objective with Stair Step is to challenge those assumptions by relying on the process structure, not the organizational structure, to select a pilot application.

The workcell will be the principle building block of a stair-step model and of the workflow definition.

Your objective with Stair Step is to challenge those assumptions by relying on the process structure, not the organizational structure, to select a pilot application.

Although much of this sounds complex, it is utterly straight-forward. The Stair Step method uses a two-dimensional matrix with workcells listed across the top and applications down the side (see example). You begin its creation by listing all of the known applications and workcells along their respective axes. Then go back to your group of twenty-five people and ask where applications and workgroups intersect. Place an x, checkmark, or a bullet at each cell where the two intersect. As you go through this process with your group, new applications and new workcells will emerge. Add these to the matrix as you continue your discussion. Keep in mind that there is a many-to-many relationship between people and workcells. (A workcell can contain many people, and a single person may belong to many workcells.) Because of this you will find, inevitably, that the number of workcells grows rapidly when you begin this discussion exercise. It will quickly reach a plateau, however. Once you are satisfied that all workcells and applications have been defined, recreate the matrix and eliminate x, check, or bullet marks.

You are now going to do something that is admittedly odd but amazingly insightful. Over the years, I have found that two things impact the complexity of a workflow application more than anything else. They are the number of people in the process and the number of documents in the process. Both of these have a direct bearing on the legacy that has been created to support the process. Every other measure, whether it be a written description of the process such as a policy and procedures manual, a physical walk through of the process, or even a System Schematic will offer only a partial view of an applications true complexity—the tip of the iceberg, if you will.

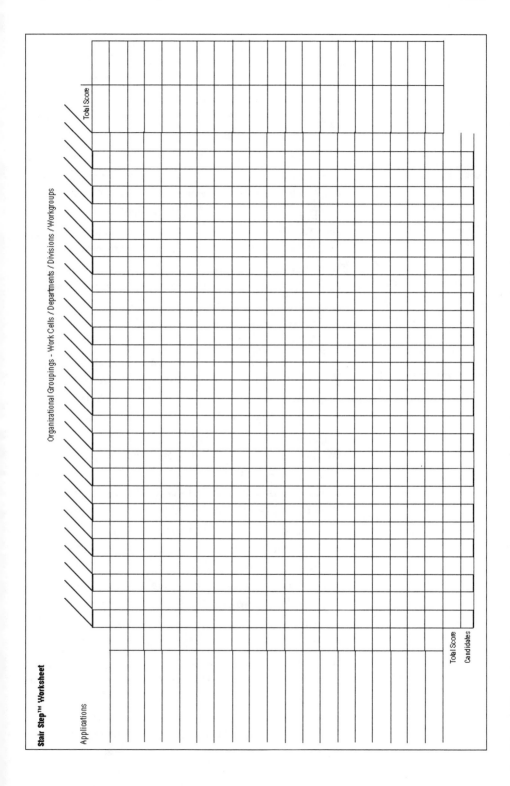

Long after the rationale behind a process is forgotten and the people who put it into place are also gone, the people who remain continue to perform the steps...

An inventory of people and documents, however, cannot hide the depth of a process's complexity. This tenet will only be palatable to you if you also buy into the premise that documents represent more than just information and people represent more than just tasks. Documents are the physical embodiment of the process and the corporate memory of the process. Long after the rationale behind a process is forgotten and the people who put it into place are also gone, the people who remain continue to perform the steps, even though they may know nothing of reasons that brought the process into being. When we interview we tend to push the people for the answers to why they do what they do. What if the answers are not in the people but in the process itself? Sound a little peculiar? This ability of inanimate things, such as document, to have a memory is a long established fact. It is the same phenomenon that is used to explain why the flow of traffic on a busy freeway is stalled long after the accident that originally brought traffic to a halt has been removed. In fact, if you watch traffic patterns from high above a freeway, you will notice how bottlenecks that appear out of nowhere will remain for some time before they disperse. In the same way processes tend to retain certain patterns although their cause has long since passed.

So how does the Stair Step method facilitate the discovery of these patterns? If you create two separate versions of the Stair Step matrix and fill out one with the number of people involved in the intersection of each workcell with each application, then do the same using the number of documents, you have an instantaneous analysis of the complexity each application's workflow entails simply by tallying the rows and columns of the Stair Step model. This tells you two things. First, if a particular workcell column's total is especially high, it is likely that the column represents more than one workcell. In other words, this workcell carries far too great a burden relative to all other workcells, and it may be that your analysis overlooked another workcell grouping. On the other hand, if a workcell column's total appears to be significantly lower than all others, it may be that this workcell should be absorbed as part of an existing workcell. Keep in mind that at this point you have not yet begun to substantially re-engineer the process. That will come with using Time-based Analysis, which looks at tasks. Tasks represent a much lower level of detail than applications.

The *relative* complexity of each application and workcell, that these totals show, forms a sort of internal benchmark. If you average these totals across applications, your ideal pilot will almost always be found within one standard deviation of this average for both Stair Step models (document- and people-based).

The Stair Step model is used not only as a selection criteria for the pilot, but also as a road map for each successive application. The first application, or step, is merely an excuse for employing ongoing change along an interminable stairway. Imagine an Escher print with a stairway that winds back onto itself in a never-ending loop, and you begin to get the gist of the Stair Step process. You are no doubt wondering whether this Mobius strip change process is merely change for the sake of change. Absolutely not. Stair Step does not cause the change, neither does it stem it. Stair Step simply recognizes the volatility of and the constant struggle to adapt to a changing organization. Change is brought on by external forces. Think back to the discussion at the beginning of the book about the constant acceleration in the creation of new competitive

Imagine an Escher print with a stairway that winds back onto itself in a never-ending loop, and you begin to get the gist of the Stair Step process.

benchmarks. If you are caught in an unceasingly changing organization you need tools with which to analyze the nature of that change and the effects it has on your workflow.

Completing the Model

In the completed Stair Step model, the pilot applications is adjacent to other applications and workcells with which it shares certain common cells. The degree of commonality can be used to determine which applications will be best leveraged from the effort put into the pilot application. The same can then be said of the third, fourth, fifth (and so on) applications. Each leverages the efforts of its predecessors. We refer to this process as that of *affinity analysis*; determining the affinity between workcells and applications is one way of streamlining not only your work processes but also the process of automating them.

It is important, to re-emphasize the necessity to keep the Stair Step current. If you create a Stair Step model and then put it on the shelf for even six months, chances are good that some of the workcells have already changed. This may, in turn, have an effect on your affinity analysis, the relationship of one workcell or application, to the next. Indeed, the very nature of implementing a single step will often be cause to re-evaluate the next steps in implementation.

Now that you have a sense of how Stair Step works, let us go back to our description of Stair Step as a tactical approach to change. Although the description of Stair Step may appear tactical and contrary to the popular belief in re-engineering at large in order to achieve maximum benefits, it actually takes into account the entire organization and all of the applications that fall under our sponsors domain. Instead of attempting to reverse-engineer and rebuild the enterprise from ground zero, which may not be practical given the competitive pressures and ongoing business requirements of the organization, Stair Step provides an enterprise learning curve that builds on previous success and experience. The end result is an enterprise system that addresses the impossible-to-predict nuances of individual workgroups and applications, while meeting the evolving needs of the entire enterprise.

This method is very successful with workflow because it encourages early appreciation for the impact and benefits that the new technology will have. These benefits are often difficult to measure and appreciate until an application is actually implemented.

The Stair Step Method also sets the stage for on-going modification and fine-tuning of the system. Because the method tackles one manageable piece of the system at a time, modifications are easier to make. The Stair Step Method has another less obvious advantage, in that it provides a source of information about the ways in which the system will be used at each step. This will prove to be a valuable knowledge base upon which to build each iteration of future applications.

What Stair Step does *not* do is provide a template for re-engineering a process. Nor is it a replacement for wholesale re-engineering in the case of organizations that are in crisis mode and therefore must drastically alter their business model substantially in order to simply survive. Again, though, here is the irony; without a Stair Step process in place, an organization is doomed to face crisis after crisis as it finds itself in a perpetual cycle of obsolete business practices, work processes, and computing infrastructure due to the inability to constantly adjust to change.

Without a Stair Step process in place, an organization is doomed to face crisis after crisis as it finds itself in a perpetual cycle of obsolete business practices.

This is somewhat like learning to drive an automobile. One of the subtle aspects of driving is the constant minute adjustments a driver makes to follow the path of the road. The response of moving the steering wheel up to 180 times a minute is so natural that it can be termed almost an autonomic response. Driving students, unaware of this phenomenon, find themselves heading directly for the curb and as a result making enormous, sudden, and heart-stopping adjustments to correct their course and avoid a collision. Luckily, a collision crisis happens with such a high frequency, that after only a few hours of instruction most people learn this necessary skill.

Crises in an organizational setting are, unfortunately, spaced so far apart that they do not have the same reinforcing effect.

You could make a case that there is nothing wrong with crises management, as long as the crisis occur with regularity and not once each decade. The tremendous streamlining of organizations being experienced today, however, is creating a constant crisis mode. Viewing the organization as a constant unceasing series of minute adjustments is the only way to avoid swerving into the curbstone.

Stair Step offers a means for identifying the adjustments you need to make, however, it is simply a divining rod that points to the most likely application of change management and its most likely successors, at any given point in time.

Using just the two methods described so far, the System Schematic and Stair Step, you may be able to point to the areas where change is most likely to help streamline your processes, but you would still have no way of analyzing the specific flaws in your processes or the payback in changing these processes. This will require the third, and most powerful, method in our analytical arsenal, Time-based Analysis.

Time-based Analysis: Collapsing the Business Cycle

Organizations are made up of a series of intricately intertwined business cycles. When considering how to streamline an organization, through the use of workflow, business cycles are the first thing at which to look. The objective of workflow analysis is to redefine and then reconstruct the components of lengthy business cycles in such a way that the time required to execute a task is minimized and the transfer time between tasks is eliminated entirely.

For example, a major West Coast banking institution needed to streamline a policies and procedures application. Upon examination of the existing system, it became apparent that new policies and procedures sometimes required more than one month to be approved. When the tasks that made up the approval process were listed, however, they totaled only five days. The business cycles, on the other hand, averaged fifteen days. The discrepancy was in the timing of the monthly Board of Directors meeting held on the second Tuesday of each month. At these meetings, every new policy and procedure had to be read into the minutes and approved by the board. The result was a process bogged down by the politics of the organization.

A process retains the memory of its form even after the individuals who first gave it form have departed.

Processes steeped in history are not atypical. As we have already discussed, a process retains the memory of its form even after the individuals who first gave it form have departed.[3] Because, business cycles are often associated with obscure organizational habits, culture, and politics, the solution to uncovering bottlenecks and redesigning the process must include overcoming organizational obstacles. The application of workflow technology

---

[3] Refer to the earlier discussion on page 131 about process memory.

is a secondary issue that may, in some cases, not even be required once the business cycle is analyzed and redefined. It is important to understand and to appreciate this when assessing the organization's workflow and recommending technologies to help streamline it.

In another example, a major East Coast bank reviewed its credit card approval process, which required an average of ten business days from submission of a new application to approval or denial. The actual time spent in the work of processing the application was two hours. In other words, 99.75% of the business cycle was dedicated to *transfer time* and not *task time*.

This astounds many people when they first hear it, but not as much as the fact that the standard ratio of transfer time to task time, across almost all industries, is nine units of transfer time to each unit of task time. There is another, more interesting way to state this that makes sense to many frustrated office workers. 90% of work gets done in the last 10% of time allocated for the task![4] One way that we attempt to correct this is to create constant tension in the organization to get work done—effectively, we, make every minute seem like the last ten. This is pure management by fear. It may work for a while but ultimately it results in burnout and poor quality.

Consider carefully what the nine-to-one ratio means. What probably strikes you at first is that, if not for transfer time, organizations could conceivably increase their throughput by ten-fold, producing more product with less overhead. That is not the benefit of eliminating transfer time. If it was, we would have decimated transfer time a long time ago.

---

[4] Quality Digest, May 1994, Robert P. Reid, There's more to quality management than TQM.

Until very recently increasing output provided a competitive advantage. The manufacturing techniques and substantive advances in productivity during this century came from the ability to produce more goods to be consumed by a growing global social affluence. As we have already seen that equation has begun to shift over the past three decades. Remember that, at present, the world can produce more hard goods (cars, stereos, etc.) than the world can consume. Production has clearly outstripped demand.

So, where is competitive advantage to be found if not in the ability to produce more? Higher quality is certainly one avenue that has been exploited with almost fanatical zeal, but even this playing field is quickly leveling. The winners are not those who can produce more, but those who can innovate more and faster than their competitors.

A McKinsey study shows that, on average, companies lose 33% of after-tax profit when they ship a product to market six months late, as opposed to only 3.5% if they spend 50% more on the development than planned but then ship on time.[5]

The ability to quickly innovate new products and continuously outperform competitors' time to market is a bastion of competitive advantage, especially when transfer time takes up at least 90% of the time in the innovation, production, and delivery cycle.

The dependence of today's enterprise on fast innovation and delivery of product can not be overestimated. Consider that 50% of Hewlett-Packard's sales come from products introduced in

The dependence of today's enterprise on fast innovation and delivery of product can not be overestimated.

[5] House, Charles H. and Raymond L. Price. *The Return Map: Tracking Product Teams.* HBR. February, 1991.

the last three years. The automation of the innovation cycle is becoming the twenty-first century equivalent of automating production cycles during the twentieth century.

What we fail to recall, however, is that the industrial revolution began over two hundred years ago with the textile industry and use of steam power. In the two centuries hence we have developed a wealth of knowledge about manufacturing techniques, methods, and processes that enable the automation of the factory.

But what similar methods exist for the automation of office processes? Even Frederick Taylor's stopwatch-driven time-motion studies were intended for rigorous factory-like environments and not for the highly interactive free form processes that drive today's information-rich enterprise.

What about project management techniques, CASE tools, dataflow diagramming, and systems analysis methods? Don't these qualify as business process analysis tools? Yes, absolutely, and nothing we say here will diminish the importance of relying on these tools, or others with which the reader is already familiar, as part of a thorough analysis.

However, here's the rub; whatever tool you use must be capable of two things. First, the tool must be comprehensible by the process owners. This is no different than the case of the System Schematic. If process owners are not able to readily infer efficiency or inefficiency from a visual depiction of the process, they will not be able to participate in redefining it. Second, the tool must be capable of analyzing three types of process time: Task Time, Transfer Time, and Queue Time. Amazingly, some tools, such as dataflow diagramming methods, completely ignore the basic element of transfer time and focus exclusively on task time. Dataflow diagrams emphasize the portion of the process computers have been able to automate prior to the broad-based availability of networked computer applications.

If business processes were completely efficient and therefore consisted only of task time (which, as we have already said, is patently false) there would be little need for Time-based Analysis. We simply could define the series of tasks that comprise a process, identify the cost of each task by multiplying the resource cost by the task time, and then associate a cost with the total value chain of activities to produce a given product or service. But any accountant or manager will tell you that is much easier said than done.

Activity-based costing in most office environments is virtually impossible. A typical office worker is involved in a variety of tasks at any one time, making the allocation of a specific resource to a specific value-chain of tasks difficult even in small organizations. The process cycle time, on the other hand, is an equally erroneous measure of the value-chain duration, since the value-add portion of time in the value chain is always a fraction of the time to complete the business cycle. (Refer to our case study regarding the credit card approval process, in which value-chain activities took less than one percent of the time required to complete the approval cycle.)

How does Time-based Analysis (TBA) help?

First, it offers a realistic perspective of the business cycle and the value chain, or value-add portion of a business cycle, by measuring not only task time but, more important, the business cycle time between people. In doing this, Time-based Analysis also offers a unique benefit for organizations that anticipate cultural resistance to analyzing their processes. Because TBA is based on the premise that 90% of the business process problem lies in the process itself, and not in the tasks

Starting with the premise that people are the problem will alienate both good and poor workers.

Time-based Analysis provides an objective measure of the existing problems.

performed by the people, it minimizes the perceived threat of the workers themselves being identified as the problem. This is not to say that human factors are not a key aspect of workflow and of re-engineering. It may be that further exploration will demonstrate that there are personnel issues that must also be considered, but starting with the premise that people are the problem will alienate both good and poor workers. Additionally, many problems with prolonged task times result from deficiencies in the current system's ability to transfer information in such a way as to optimize existing peaks and lows in users' workloads.

Organizations may look upon these as management-related problems that can be solved by increasing the efficiency of the task and worker. This approach often ignores the real problem—transfer time. In many cases simply eliminating the transfer time can substantially improve a business cycle. Add to this the ability to track the document and spontaneously interact with individuals throughout the workflow process, and workflow may change the total cycle time without changing a single task, but rather by collapsing the idle time between tasks. Think back to the discussion about process intimacy and the effect it has on process efficiency.

Second, Time-based Analysis provides an objective measure of the existing problems. Although it is possible to debate the inefficiencies of work habits, information systems, and infrastructure, the element of time is easily measured and presented. In analyzing a process for potential workflow improvements it is important to be able to present objective and quantitative metrics, rather than rely on intangible benefits that cannot be readily justified. This is true even when justifying workflow based on future opportunity.

Third, the process of defining task transfer times and total business cycle times will expose the full breadth of inconsistency and complexity of the business cycles in question. It is not unusual to find that some business cycles have no consistent time duration. In the policies and procedures example given earlier, the time required for the approval of a given policy varied widely from two weeks to six months. What appeared to be one business cycle actually consisted of multiple formal and informal cycles. Until Time-based Analysis was applied it was impossible to identify the problems with the existing process or the applications for improved workflow.

Although the formal methodology and tools for Time-based Analysis require some skill and experience and are beyond the scope of this text, the premise and concepts can be easily grasped and applied to all aspects of a workflow analysis. One of the significant differences, however, is that we are now describing individual tasks rather than applications in the left-hand column. Otherwise, there are some minor alterations to the Stair Step format, but all in all you should be able to navigate through this with relative ease if you have already made an effort to understand Stair Step.

The first step of Time-based Analysis is the definition of the tasks involved in a given business cycle and the workcells that perform work on these tasks.

The definition of the business cycle for which you perform the Time-based Analysis will correspond to the Stair Step application that you have decided to analyze. It should be broad enough to include all of the activities for a certain logically connected set of tasks that represent the application. For example, publishing a policy requires, at a minimum, several tasks: initiating the new policy, reviewing and authoring the policy, executive approval of the policy, checking its consistency with existing policies, board of directors' approval, and typesetting, publishing, and distributing the policy. Don't be overly concerned with getting every single task defined from the outset.

As with the System Schematic, some dialog and discussion in iterations is a better approach than trying to get it all right the first time. As each task is defined it is associated with a task time. Task time is the total amount of time it takes to complete the task for a given workcell, exclusive of the time spent waiting for information related to the task or subsequently transmitting information to the next task. For instance, the task time for board of directors' approval of a policy is never more than two hours, since this is the length of the directors' meeting. It may, however, take two weeks to actually approve the policy because the directors meet only twice each month. This is represented in the second variable, which is the transfer time between tasks.

Transfer time identifies the time required to transfer information from one task to another. At first this may seem a trivial distinction. After all, isn't transfer time part of the task? In the case of the directors' meeting, once the author has written the policy, it's no longer his or her job to get it approved. Instead, approval becomes the responsibility of the directors. As far as the directors are concerned, however, they fulfill their responsibility through meeting twice each month. The problem becomes painfully obvious—transfer time belongs to no one. No one takes responsibility for it and no one is to blame. On the other hand, if we arbitrarily say that it is the problem of the directors to meet more often we are suddenly putting someone in the position to defend a piece of the process that they cannot control. It's no surprise that business cycles are difficult to collapse if there is no accountability for one of the most time-consuming elements of the business cycle.

It is for this reason that we initially separate the two components of workflow analysis. Tasks clearly belong to individuals and workgroups. Transfer time, however, belongs to the process itself and the information systems in place to facilitate the process. Once we have established that transfer time is a process issue and not a people issue, it becomes much easier to identify problems with the existing process. In the example, transfer time is not associated with one or the other of the receiving or sending workcell, but rather to both. That will create a false tally if you add up all transfer times. The point, however, is not simply to look at the total of task time versus transfer time (although this does provide some insight), but rather to identify which tasks are preceded or followed by inordinately long transfer times.

It's no surprise that business cycles are difficult to collapse if there is no accountability for one of the most time-consuming elements of the business cycle.

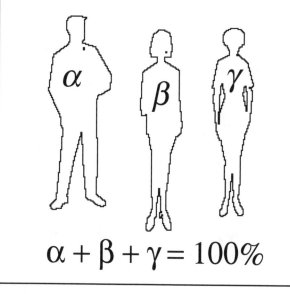

$$\alpha + \beta + \gamma = 100\%$$

The Myth of Cycle Time

Most people believe that the total time required to complete a task is equal to the total time required to complete a business cycle. This is only true if there is no transfer time in a business process.

A simple analysis of the polices and procedures approval cycle, described earlier, shows that the transfer, and queue, times were significantly greater than the actual task time involved. If the process were changed to require the Board of Directors approval only on a subset of the policies and procedures the business cycle could be reduced substantially. In fact, since 95% of all policies and procedures were approved by simply being read into the Board meeting minutes, the process was being held hostage by the two week transfer time.

By evaluating the relationship of transfer time to task time we can begin to formulate opinions as to likely workflow candidates

By evaluating the relationship of transfer time to task time we can begin to formulate opinions as to likely workflow candidates for streamlined business processes. For example, if the total transfer time is significantly higher than the task times, there is a great deal of opportunity to reduce information transfer time without necessarily changing tasks. The benefit is, of course, less disruption on the part of users. It may also be the case that transfer times vary dramatically from one set of tasks to another. This may be indicative of a bottleneck situation that occurs again and again in the same area regardless of the information being processed. That may represent a point solution for automation that again minimizes user disruption.

You may similarly be able to draw conclusions about the interaction of workcells. For example, if transfer times consistently appear to be lengthy when transferring information between two workcells, it may be the case that there is a specific situation that causes these two groups to delay transmission or receipt of information. What the problem is cannot be determined from the Time-based Analysis alone. But it is certain at this point that you will not be pointing a finger of blame at one group or the other from these results alone. Again, this is a key reason for the allocation of

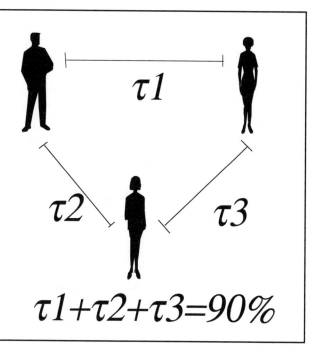

The Reality of Transfer Time

In conventional business processes, transfer time accounts for about 90% of the total business cycle. In the case of one major New York Bank, only two hours out of a ten-day business cycle were attributable to task time. That makes the percentage of transfer time an astounding 97.5%!

$$\tau 1 + \tau 2 + \tau 3 = 90\%$$

transfer time to both sender and recipient of the information.

In every case, the Time-based Analysis model can deliver a concise assessment of the existing workflow environment and thereby manifest potential areas of opportunity for a new workflow system.

Keep in mind that throughout our discussion of Time-based Analysis we never said that people are never the problem. That may well be the case, and it may have to be dealt with as part of the workflow analysis. Starting with that premise, however, is dangerous and time-consuming. The approaches presented in this chapter shift the focus to redefining the process model. Users are much more likely to cooperate in that case and the underlying inefficiencies of the business process are much more likely to be resolved. Applying Time-based Analysis will help you to demonstrate short-term payback that will provide long-term leverage for your work-flow applications.

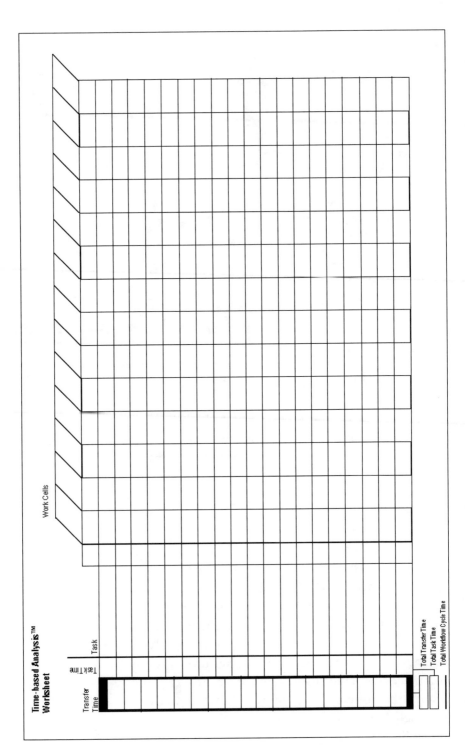

Workflow: The Technology Imperative

Technology Components of Workflow

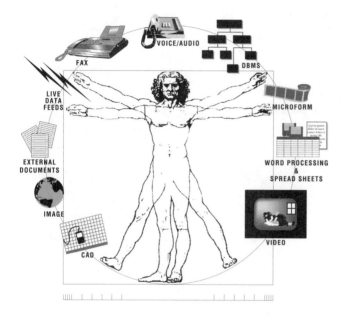

In order to understand the technology components of workflow we need to first understand its evolution. Workflow evolved as a means by which to unify and coordinate rapidly growing networking infrastructures, the resulting workgroup computing environments, and the many applications and office automation tools available to end users. Placing workflow in this context helps one to appreciate the soundness of workflow technology and its role in today's information systems.

Although workflow has evolved as a separate and distinct solution, its roots are clearly linked to several established technologies including: E-mail, project management, databases, object-oriented

151

programming, and CASE tools. With more than seventy workflow products in the market, and a market growth rate in excess of 35% yearly,[1] there are a variety of approaches combining different aspects from each of these technologies.

These diverse workflow approaches and solutions are similar however, in one fundamental manner—they all attempt to solve the problem of fragmented and isolated information. Only recently has this been possible to consider. If we consider the evolution of information systems over the past five decades, three distinct paradigms emerge. These have led to the present day model of distributed computing, shared information, and the need for workflow.[2]

- First Paradigm technologies (circa 1950s–1960s) were centered on the computer program that drove the information processing in a highly isolated model—the Von Neumann model of computing.
- Second Paradigm technologies (circa 1970s–1980s) shifted the focus to the data models in a highly centralized environment. These were typified by mainframe and large database applications.
- As we entered the Third Paradigm, (1990s) emphasis again shifted, but this time away from the technology towards business processes and the tasks comprising them.

---

[1] 1994 Study of the Workflow Market, Delphi Consulting Group, Inc.

---

[2] Koulopoulos and Frappaolo. *Electronic Document Management Systems.* McGraw-Hill. 1995.

The Third Paradigm brings full-range computing into the personal domain of individual workers at all levels of the organization. For the first time it is possible to create and easily propagate isolated computing environments throughout an enterprise.

It was the advent of Third Paradigm information systems that resulted in true "islands of information"—with virtually no connection to the rest of the organization.

At the same time, these paradigms were evolving, the nature of applications development was undergoing a radical shift as it matured through increasing levels of abstraction. Applications that required sophisticated programming techniques and dedicated applications developers, and months of development time can now be created in days or even minutes. The graphical computer development environments and graphical user interfaces that surround us today are more than the icing on the cake. They represent an entirely new recipe for applications development. The success of computing is now contingent on the ability to join individuals and workgroups with radically different computing process solutions within a single cohesive and aligned enterprise.

Incompatibility has flourished among applications and data repositories, across vertical and horizontal lines of the organization.

As these independent systems proliferated, they became vast warehouses of information. The information they contain has become the foundation for individualized empowerment and advantage. Systems can be customized to meet the needs of each user. In many cases, the users themselves can customize the systems. From a user's standpoint, diversity is a strength.

But there is a downside to this success and rapid proliferation. Incompatibility has flourished among applications and data repositories, across vertical and horizontal lines of the organization. Ironically, more paper than ever before is being generated by computer-based systems. Perhaps the greatest beneficiaries are the suppliers of printing hardware. Seven hundred million pages of computer output are generated each day, and 70% of that paper is used for data entry into other computer systems—a testimonial to the ineptitude of different computer environments to communicate. The information archipelago has arrived and the bridges from island to island are paved with paper!

Since workflow has evolved from this environment, as a result of the migration towards distributed workgroup computing and higher-level languages for the definition of process automation, it is not so much a radical new technology as it is a logical next step in the use of computing to integrate knowledge-based tasks and activities. Its primary benefit is joining islands of automation, which use discrete tool sets, into enterprise information systems.

In this sense, workflow is an operating environment that provides a context for working with other technologies. Ultimately that trend will lead to workflow's incorporation as a standard component of operating systems, but as that happens the concept of an operating system will itself evolve dramatically. Current business processes are 90% disparate and fragmented. When a *Business Operating System* (BOS) evolves, it will bring together, use, and finally disguise the traditional operating systems of the last few decades in the same way as today's high-level graphical user interfaces cloak the complexity of programming languages and application development environments.

Workflow is the first substantial step in uniting process across an enterprise and the evolution of the Business Operating System. Understanding the nature of this Third Paradigm, the genesis of workflow, and the advent of the Business Operating System we can look at the technology components of workflow, understanding that workflow is not a singular application but rather the orchestration of many applications, information sources, and processes.

Types of Workflow

Workflow technology can be applied to an organization using any combination of four development methods; ad hoc workflow, transaction-based workflow, object-oriented workflow, or knowledge-based workflow, and three process models; mail/message-centric, document-centric, and process-centric. The intersection of each development method and process model is intended for a specific type of application. (See chart on page 161.)

Ad Hoc Workflow

Ad hoc workflow applications are intended for use by dynamic workgroups that need to execute highly individualized processes for each document. Although some reusability of a process model (the rules that make up a workflow) may occur, there is not enough consistency to justify the effort of creating structured transactions. Alternatively, the process may be so volatile that a definition of any standard rule set is considered to be futile. Ad hoc workflow requires the use of graphical workflow development tools that are easily created and modified by the end user.

One type of ad hoc workflow that shows signs of significant promise is that of E-mail-enabled workflow. E-mail-enabled workflow uses either existing E-mail applications as the messaging standard for transmitting workflow documents or a proprietary E-mail environment that is provided with the workflow application. One of the benefits of this approach is the simple and comfortable metaphor that E-mail already has established in most organizations. This minimizes the cultural resistance and investment in customized application development. On the other hand, it is not always suitable for large transaction-based workflow applications, which require highly customized applications.

Transaction-based Workflow

Transaction-based systems are intended for highly structured applications that include lengthy and complex tasks. The rules that define a workflow can be precisely defined and executed under a transaction-based model. (All workflow is based on rules, which is explained in this chapter.) What is unique about transaction-based workflow is a generally high volume, production-based environment that utilizes a process that entails repeated tasks and little variation among cases. In this type of a static environment, throughput is the primary concern. Documents take similar paths through the system each time and workflow simply helps to ensure the integrity of the process.

Object-Oriented Workflow

As the workflow market matures, users of workflow software are beginning to push the envelope of existing technology by exhibiting increased interest in large-scale enterprise applications of workflow. This new wave of workflow applications has prompted a flurry of interest in an old subject, with a new twist—object-oriented technology. The trend is significant as the workflow industry moves further into the ranks of application development software and closer to the mainstream of information systems. That signifies what may be a fifth generation of applications development tools based on an object-oriented workflow metaphor. Without any standards or benchmarks in this incipient industry, however, each vendor is adopting its own approach to object-orientation. That creates the potential for even more confusion in an already schizophrenic market space. There are, however, some basic features and similarities among this new cadre of products that evaluators can look for when assessing the viability of this new approach to their workflow applications.

Object-Oriented Workflow Combines Documents and Rules

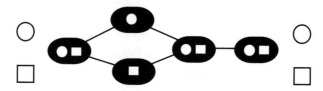

In an object-oriented workflow application, two instances of a workflow (circle and square) represent the same basic workflow rules and process but take slightly different routes through the same process.

Elements of Object-Oriented Workflow

The purpose of object-oriented technology, is to enhance the developer's ability to create complex applications, to increase the integrity of these applications, and to create an interface for both developer and user that is easy to navigate and use. Object-oriented workflow technology accomplishes this by providing some basic features such as encapsulation, inheritance, referential integrity, nested procedures, procedure libraries for enterprise application development, and icon-based graphical environments for the development of workflow applications. The most visible component of object-oriented workflow is this last facility, the graphic development environment. A variety of vendors have already adopted the use of iconic workflow desktops. Developers, using these environments, manipulate icons that represent tasks, users, processes, and routing objects. The objects can be dragged and dropped from the palette to create a visual representation of the workflow process model. This alone does not, however, create an object-oriented approach. For example, some vendors go further to provide integrity across applications that share the same library procedures. In this case, a change made to a low-level procedure is reflected in all instances where it is used.

Workflow objects consist of both information and process knowledge.

It is important to understand that workflow objects consist of both information and process knowledge. Both are combined to create an "instance" of a workflow. The benefit of this is that each instance can follow a separate route through a business process, although the same basic workflow model is being used. For example, in a mortgage processing application, loan forms are routed through the workflow based on certain established rules. If these rules change due to a change in federal regulation, then new documents have to follow a new workflow process. Without an object-oriented approach however, we would have to wait until all current work in progress is completed before changing the workflow for new work. Object-oriented instances are not affected by this because each document is a self-contained object complete with its own set of workflow rules. By the same token, these objects can be modified while in process without affecting other objects in process. This is an especially critical facility for workflow applications that rely on a combination of production and ad hoc workflow.

Another important element of object-oriented workflow applications is that of the development environment. In an object-oriented approach, application integrity increases and development time decreases. In large part this is due to the use of procedure libraries. These act as reusable workflow building blocks. In a complex enterprise application each procedure may be used in hundreds of workflow process models. (The process model is the highest level of definition in a workflow.) When a change is made to the procedure it is automatically propagated throughout all process models that also use the procedure. Object-oriented products use this type of approach to manage the ongoing volatile nature of many workflow development environments.

In the case of some products, the scripting language or API sets used to define the workflow are being bundled, or *encapsulated*, into object libraries. These are often GUI-based development environments that can be recombined to create workflow applications. They are not end user environments, but rather highly abstracted developers tool kits, that enable quick application development. If existing objects are not sufficient to create the desired workflow application, new objects can be developed and added to the library. In this way enterprise application libraries are created over time—representing a valuable knowledge-base.

Although the trend towards object-oriented workflow is important, it is equally important to recognize that much of the workflow technology in use today does not fall into the strict definition of object-oriented. Instead, there is a migration of workflow into development environments that are better managed through the use of object-oriented techniques. If your evaluation calls for a demanding application development environment, it is probably worthwhile to consider candidate vendors with object-orientation as one important criteria.

Components of Object-Oriented Workflow

Object-oriented workflow incorporates several key features of traditional object-oriented technology and databases to enhance the ease of development and to increase the integrity of workflow applications.

- Object-oriented—A system that combines information and processing rules in a single object.
- Encapsulation—The ability to combine multiple procedures into a single procedural representation. This is most often done through the use of an icon-based front-end.
- Procedural Nesting—The ability to embed complex procedures within other procedures. Encapsulation is often part of this, as is the ability to drill down through multiple nested levels.

Types of Workflow

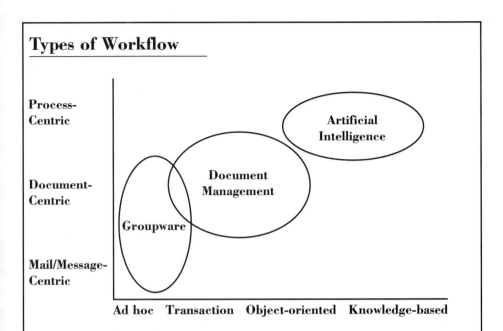

Process-Centric

Document-Centric

Mail/Message-Centric

Ad hoc   Transaction   Object-oriented   Knowledge-based

The model depicted above defines the market from the perspective of both the development environment and the process model. The development environment is represented by four types of workflow applications: ad hoc, transaction-based, object-oriented, and knowledge-based workflow. The process model is represented by three distinct categories: mail/message-centric, document-centric, and process-centric.

Some products claim to be suitable for all types of workflow, but evaluators of workflow technology should be wary of this claim. Although products may offer some level of functionality suitable for each type of workflow model, they will not provide the full range of capability necessary for the successful implementation of all types of workflow.

While all workflow applications provide for the automation of business processes, an organization should have a clear idea of the orientation of its workflow application before product evaluation begins, as the selection of the appropriate software product will be based in large measure on the type of workflow application required.

- Instances—A workflow process for a particular information object. Instances act as individualized workflows for each document processed through a workflow process model.
- Procedural Library—A collection of procedures that can be recombined and reused in such a way that a change to the procedure in the library is automatically incorporated into all other process models which rely on that procedure. This is also often called referential integrity since procedures referencing other procedures are always up-to-date.
- Inheritance—The ability to create new procedures that are based on the rules and properties of prior procedures as their baseline.

A workflow object acts as an envelope. It contains not only the information being routed but also the knowledge about its path through a process, the roles that act on it, and the nature of the tasks that make up the workflow.

Knowledge-based Workflow

Knowledge-based workflow goes beyond the simple rules-based approach to provide methods of incorporating exception processing. This may be done through training methods that use other tools such as artificial intelligence (AI) or expert systems. This would not be dissimilar to the concept of training a computer system to recognize new patterns by using a neural network or an inference engine. In a knowledge-based workflow system the workflow would use statistical, heuristic, and artificial intelligence to infer correct routing, scheduling, and exception routing. This would alleviate the problems inherent in anticipating every rule and variable that may impact a business process.

Although no vendors support knowledge-based workflow at the time of this writing, it is one of the most interesting and significant developments for workflow. Many vendors have already expressed an interest in this type of workflow. Commercial applications would include complex business processes that cannot be defined through analysis due to the difficulty of collecting all of the rules that govern dynamic business process. It would also address processes that are subject to variables which change in unpredictable ways. Products currently claiming to be knowledge-based are using less sophisticated transaction-based methods.

Mail-Centric

Products in this category emphasize the benefits of using existing modes of electronic communication, primarily E-mail, as the messaging and delivery service for a workflow system. Many products are also bundled with forms-based utilities which provide screen definition of forms for a variety of uses such as routing and approvals. Many products in this category also are beginning to express a strong interest in the concept of intelligent agents, which can perform workflow tasks even if the individual responsible for a task is unavailable.

Document-Centric

These products focus on the document as the unifying object in a workflow process. Documents are associated with owners, applications, and rules that govern their routing and processing. Many document-centric workflow products also provide facilities for document management, such as check-in/check-out features and redlining. Others expand on traditional metaphors, such as the filing of documents in file folders, to associate processing rules with folders.

Process-Centric

The majority of high-end workflow applications use process-centric products. These products often rely on an underlying database, such as Oracle, Sybase, or Informix, to store workflow data and definitions. They also provide extensive programming facilities for the scripting of workflow applications. In many ways, these tools resemble high-level programming languages. The difference is that they do not impose new end user tools for the development of a workflow application. Instead they act as an underlying facility for the management of an organization's workflow.

Workflow Architecture

Workflow applications use five distinct layers of structure to categorize the organization of information management activities: Processes, Cases, Folders, Rules/Applications, and Documents. In addition, the rules are defined in terms of Roles, Routing, and Work Queues.

Processes

The process model, the highest layer, consists of a series of tasks and rules that must be defined for each process that takes place. For example, an accounts payable process is made up of several tasks, such as the receipt of the invoice, review and approval, reconciliation with the receiving report, etc. Specific tasks have rules associated with them.

The Workflow Architecture

The architecture of a workflow system includes at least four layers of information, in addition to the Roles and Routing information.

Folders

Roles
Routing

Data
Documents

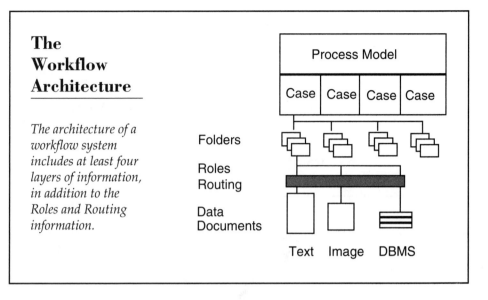

For example, invoices over a certain dollar amount require a higher level of approval. Each task and its associated rules must be defined to the workflow system. We will explain later how the rules layer of this model is further broken out into a series of attributes that define the way in which documents interact with the process and folder layers.

Cases

The next layer is the case, which is an individual occurrence, or instance, of the process model. The processing of a single invoice for payment would be considered a case in the accounts payable process model. Each time the workflow procedure is invoked, a new case is created. Cases are also referred to as *instances* by some workflow vendors. The terms can be used synonymously.

Folders

The third layer is the folder which contains a logical group of documents. The folder may contain any combination of data types including text, image, and data from multiple sources. A folder for an accounts payable application may contain the purchase order, receiving report, and financial data from the general accounting system.

Folders

Folders often act as the repository for all of the objects contained in a workflow process. This includes information and, in many cases, rules.

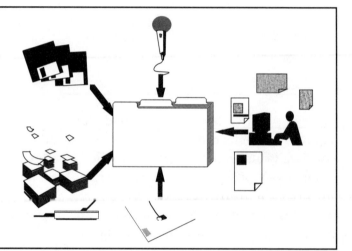

Rules

The fourth layer consists of the rules associated with the processing of documents. These rules define the specific processing activities that are involved in routing workflow documents and working within specific applications. For example, a particular contract document may be associated with a schedule of approvals. The approvals are governed by rules that relate to the amount of the contract. In addition, an approval may in turn launch an application that updates a spreadsheet cell.

Roles

The definition of a specific workflow requires the establishment of roles for all participants. Each participant has established roles which must be defined as part of the workflow definition. Each participant is described in terms of location, job function, supervisor, and security level. An employee may be an individual participant as well as part of a workgroup. Workgroups can consist of a group of individuals working on a project, a department, or a group of individuals that share the same job function. One individual may belong to multiple workgroups at one time, depending upon the type of application. Role definition enables the workflow system to distribute tasks to the appropriate individuals for completion. It also allows for workload balancing and distribution to the appropriate employee based on skill level.

Routing

Workflow routing controls how documents move from point to point in the workflow. There are three types of workflow routing scenarios; Sequential, Parallel, and Dynamic or Conditional Routing.

Sequential Routing follows a linear path from one task to another. It is clearly defined with little variation. One task must be completed before the work is routed to the next point.

Parallel Routing enables multiple tasks to occur at the same time. In an accounts payable application, for example, a document may require the approval of three different individuals. Since the three approvals are independent of one another, they can occur simultaneously. Eventually all three are brought together at a rendezvous point. They are held there until all approvals are obtained before initiating the next task. It should be pointed out that Parallel and Conditional Routing are not necessarily synonymous. In a parallel workflow, routes may be completed at different times. Concurrent workflows are completed at the same time. Although both come together at the same point, the parallel process may include idle capacity along one of its routes.

Conditional Routing is determined by conditions which occur dynamically in the process. The system must be able to determine the appropriate route based upon the information that is received along the way. For example, an accounts payable invoice over $10,000 requires an approval of a vice president in addition to the department director. Upon receipt, the workflow system incorporates conditional routing which would automatically route all invoices over $10,000 to the appropriate vice president for approval.

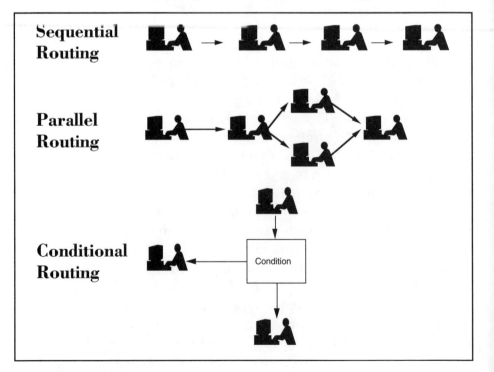

Data and Documents

At the lowest level is the data itself. This is represented by a single document or a collection of documents, which are enclosed within a folder, then in turn are controlled by a case. It should be noted, however, that the document is almost always associated with an application which is initiated in order to view and work with the data.

The concept of a document may appear foreign to readers who are familiar with "data" processing applications, where *data* replaces the use of the term *document*. In a workflow application, however, the document is the unifying force for all data and applications used to process data. The document represents not only the data, but also the formatting, processing, and presentation for the data.

An example of this is the processing of an accounts payable invoice through a workflow application. The invoice may require viewing of a purchase order and relevant financial reports. Therefore, the workflow system must be able to launch the appropriate programs to enable use of the information. The ability to invoke and integrate the presentation of these documents is a key aspect of workflow.

Although the specific terms used to describe these layers may vary from product to product, the concept remains the same across virtually all workflow technology. Tools are provided to enable the workflow technology to interface with the user, platforms, and data structures at each layer.

Workflow Tools

Although the specific tools required to create these environments are easy-to-use, the complexity of enterprise workflow applications still requires the involvement of information systems professionals.

Workflow applications have been written using a variety of tools ranging from a scripting language similar to a fourth-generation language (4GL) to an object-oriented CASE-like method that uses an icon-based graphical user interface (GUI). Many older workflow scripting languages were complex and required a fair amount of training and experience with the tools and command sets. Early versions of FileNet's WorkFlo and Staffware's Staffware are examples of this type of product. (Both of these products now have advanced graphical languages for applications development.) Most products available today abstract programming scripts into a series of objects which can be modified with forms-based entry or graphical models. This may limit the overall flexibility in highly customized situations, but it is much simpler for the development of immediate solutions by less-than-expert developers.

It is important to keep in mind that many business processes are very complex, and even with easy-to-use tools for workflow development, end users will not be able to develop enterprise-wide workflow applications. This is one of the widespread misconceptions about many new graphical workflow tools. Although the specific tools required to create these environments are easy-to-use, the complexity of designing, implementing, and maintaining enterprise workflow applications still requires the involvement of information systems professionals.

Workflow Scripting Languages

Workflow scripting languages are used as a method of writing code to define processes, rules, and operations for the workflow application. Object-oriented and 4GL languages are the most common scripting tools used to design workflow scripts. These tools generally are used by professional developers to design and generate code. Although easier to use than older scripting languages, they are not easily modified by end users. Workflow scripting languages are most commonly used in transaction-based workflow applications.

Graphical Workflow Editors

Graphical workflow editors are used by both professional designers and end users to develop workflow applications. These products use a palette of graphical tools that guide the user through the forms-based or menu-based dialogs that create on-screen flow charts, which depict the workflow process. This icon-based approach is easier to use than workflow scripting languages, as it relies on familiar metaphors for windowing environments, such as point-and-shoot, click-and-drag, and radio buttons.

Workflow applications developed using graphical editors are easier to modify, providing the end user with greater flexibility in development and modification of workflow applications, without the need to rely on professional developers. These advanced interfaces may, however, require additional programming in order to adequately integrate with other office automation technologies. Graphical editors are used in all workflow types but are most commonly found in ad hoc applications, in which there is a high reliance on the end user for development of individualized workflows.

Workflow APIs

Workflow Application Program Interfaces (APIs) are provided with most workflow products. They are essential for the integration of workflow with other business applications. The critical aspect of the API goes beyond the simple sharing of information. The API must be able to ensure the integrity of the information that is accessed and transmitted.

A workflow application intended to integrate a spreadsheet tool and a customized management report must be able to share data. Changes made to the spreadsheet that have an effect on the management report must be reflected in the report. Conversely, changes made in the management report that affect the spreadsheet must be reflected there as well. This type of dynamic linkage is important to the workflow environment, because common data elements are shared by a variety of applications across the enterprise. Workflow tools that manage this data can have a dramatic impact on the workflow system's data-sharing ability.

Workflow Simulation Tools

Workflow simulation tools allow designers to test workflow processes and to identify problems with the process before the workflow application is implemented. At the present time, very few workflow systems provide these facilities as part of the workflow product. Many developers rely on third-party vendors who have software simulation tools suitable for business modeling. Public domain tools such as Integrated Definition (IDEF), which is used extensively by government, can also be used to assist in this process. Although this is an area of interest for many workflow vendors, these tools are unlikely to be widely available any time soon.

Databases

In addition to workflow development tools, some workflow products rely on databases for the storage of routing instructions, procedures, and rules. Databases also are used to track the status of workflow processes and to maintain a historical audit trail of each transaction. It is here that many workflow products fall short. The database and the concept of a data dictionary are not yet fully developed in these workflow systems. (The data dictionary provides a singular definition of the attributes associated with each data item or document. The variety of sources of documents and data elements in a workflow system is so robust that it may be next to impossible to impose a singular data dictionary.)

Most workflow products use either standard third-party database products or incorporate their own proprietary databases. Integration with external databases is typically provided by using existing APIs that call external databases, such as Oracle, Sybase, or other standard database management system (DBMS) software.

In some cases, proprietary databases provided by vendors cannot be accessed as tables or connected to an external DBMS. These approaches can hinder the implementation of workflow as a user environment and result in the restrictive use of workflow in isolated workgroups or simply as a document management tool for transferring and tracking.

This is not a death knell for the technology. It is still advantageous to begin joining workgroups together in advance of enterprise applications. But problems can arise when you attempt to create larger systems that are intended to join all elements of an enterprise's office automation technologies.

Recent workflow products have begun using object-oriented databases. Not only is object-oriented technology ideally suited for workflow, as we stated earlier in this chapter, but it also offers an ideal hook to object-oriented databases. Each object in an object-oriented database can be defined and maintained independently of the others and each is defined through a series of procedures and data elements. Unlike the way structured relational databases behave, object-oriented database objects interact in multitudinous ways. Object-oriented databases simulate real-world activities and can be easily manipulated and modified. Object-oriented functions, such as inheritance and object classes, enhance workflow applications by avoiding the repetitive creation of objects and by preserving an object's attributes throughout all process models. The concept of object classes allows for the creation of a structured hierarchy of workflow objects into super classes and subclasses. For example, purchase orders may be a subclass of the procurement process model while at the same time being a super class for order processing. Encapsulation masks all of this complexity within the concept of a single object or icon that can be bundled with any combination of other objects for incorporation into a new workflow.

Workflow technology should be tightly integrated with enterprise-wide database technology. The issue of referential integrity among all of the documents and data sets must also be considered. Referential integrity ensures that changes made within one application are reflected in all other applications that also use the data. Administration should also be considered, as it will be an important part of constructing and maintaining this new environment.

But the database integration has other benefits as well. If a multimedia, or Binary Large OBject (BLOB), database is used, it will provide ready access to a broad variety of data types, including documents of all forms.

The Time-based Event-driven Model

There are three basic components of every workflow environment: events, time, and objects. Workflow technology is founded on the principle of coordinating events within an optimal time frame by defining data objects that recognize specific rules. Another way to look at this reflects the fact that events and time are the basic building blocks of a workflow system. Creating these objects, however, requires a set of tools for defining each one and the relationships among them. Here too, the availability of actual product functionality and the requirements of workflow developers may differ considerably.

Current workflow implementations use procedural methods for accomplishing the definition of events and duration. For example, the product may provide an object-oriented definition language for creating a task. The task is represented by an icon that, in turn, has been defined as a set of rules for the task. These may be the approvals required for a particular document or the integrity controls for references to external data sources. The object will also contain information about the time line that the document should travel, such as that allowed for approval before the document is rerouted to another individual.

An effective workflow tool set should provide support for object creation, management, modification, and linkage in much the same way as object-oriented programming tools do this. Functions such as inheritance, object classes, and encapsulation also should be supported by the workflow system, in order to avoid repetitive creation of objects and to preserve an object's attributes throughout all process models in which it is used.

Each object has six of these associated attributes that reflect activities inherent to every information-based activity or business cycle. These six attributes are: Initiation, Notification, Iteration, Completion, Dependence, and Duration. Each attribute exists in any transfer of information that takes part in a value-added process, a single task in a workflow cycle, also called an Event.

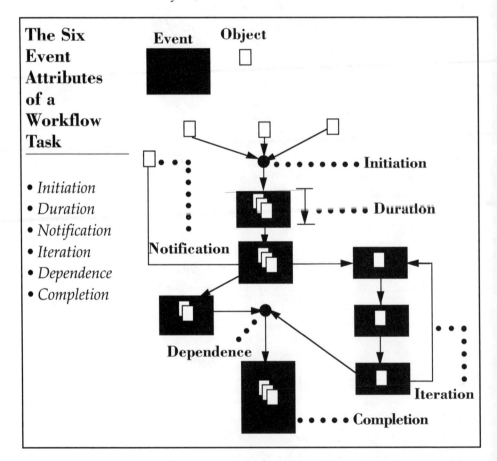

The Six Event Attributes of a Workflow Task

- *Initiation*
- *Duration*
- *Notification*
- *Iteration*
- *Dependence*
- *Completion*

- Initiation (also called a *trigger*)—An event that begins a workflow process. This may be as simple as an individual logging into workflow environment or the completion of several concurrent tasks.

- Notification—An electronic message that is initiated through the occurrence of any other workflow event. Notification may act as a trigger but is usually thought of as an end-step in the process.
- Iteration or Negotiation—The repetitive execution of an object, using substantially the same rules each time. For example, a purchase order that must be signed by three managers is an iteration of the same object.
- Duration—The time required to complete an event or the period of time for the deferment of a specific event until its dependencies have been satisfied. Duration does not include transfer time as, in a workflow system information transfer time is eliminated and only task time remains. This means that events are always queued up and waiting for information to process. It may, however, include the time specified within which the next event must be processed.
- Dependence—This represents prerequisite objects that must initiate the dependent object. In some cases there may be numerous prerequisites to an object. For example, the approval of a purchase order by three managers before an order is initiated.
- Completion—This is the last event in a workflow process for a particular object. The only exception is notification which may proceed as a result of completion. In a string of objects, completion of one object may lead to initiation of another.

Intelligent Documents: Another Approach to Workflow

The object-oriented approach can also be used by incorporating objects into a document. This approach does not provide all of the functionality of an overall workflow environment that incorporates multiple office automation applications, but it does work well within a pure document management scenario. Intelligent documents use a combination of compound document structures and rules-based routing, processing, and modification. This may be ideal for a publishing or authoring application that involves workgroups.

In these applications, the document must be transportable across multiple platforms. This is accomplished through adherence to document interchange standards such as ODA/ODIF or newer approaches such as those provided through electronic viewers. Unfortunately, electronic viewers do not address document objects but only the document layout. In both cases, however, the rules and linkages that make up the document must be transported with the document from one platform to another. The workflow system may need to work with individual document objects. Therefore, workflow system must be integrated with the underlying document standards.

The concept of intelligent documents also applies to the broader process of linking applications through embedded tools for launching applications such as Microsoft Windows' Dynamic Data Exchange (DDE) and Object Linking and Embedding (OLE).[3] DDE, OLE, and Apple's Publish/Subscribe function provide environments

---

[3] OLE is one of several approaches to creating compound documents. Others are reviewed in the section on standards.

by which to attach data and objects to the applications that created them. These are useful functions for integrating applications and information. They do not, however, take the place of workflow, which also associates the aforementioned attributes with the rules that govern the information object.

Undoubtedly, workflow will slowly move into the realm of intelligent objects and documents. This is, however, a longer term direction. In the short-term, workflow will evolve in several areas. Most noticeably lacking in current-generation workflow products is the ability to simulate workflow procedures prior to implementation. Vendors will begin providing these tools as a necessary part of the business process redesign effort of which workflow is often a part. There will also be development in the area of linking objects within workflow applications through system-level utilities such as OLE. Ultimately, this will lead to workflow's incorporation into operating systems such as future object-oriented versions of Microsoft Windows. This is especially critical, as workflow is used to provide a metaphor for the varied office automation environments currently in use. The most dramatic shift will likely occur as workflow becomes the enabling technology for business process design. It will be the conduit that enables many predecessor technologies, those that promised increased productivity, to finally deliver on that promise.

Evaluating Workflow

Identifying the correct workflow solution for your organization may well be the single most important aspect of an effective workflow implementation. As with most new technologies, the market for workflow is characterized by numerous product offerings and approaches. Trying to make sense of this chaos is haphazard at best, unless you begin to categorize the types of workflow products available and then prioritize their functionality vis-a-vis your organization's needs and requirements. That alone may be a monumental task.

This chapter is intended to assist the user in evaluating available workflow products by providing a listing of key workflow components that each product should be gauged against. Each item is described in order to familiarize you with the mechanics and the benefits of alternative approaches.

Keep in mind that not every item on the checklist will be found in each product, nor will every item be required by your organization. But each should be carefully understood and prioritized before selecting the approach that offers the greatest breadth of required functionality, flexibility, and potential for success.

Workflow is often associated with imaging. The two are, however, completely independent technologies.

Imaging Capability

Workflow is often associated with imaging. The two are, however, completely independent technologies. The dependency seems to exist due to the synergy between the two technologies when imaging is the driving application. In fact, it makes little sense to consider imaging without also considering workflow technology. The inverse is not always the case. There are many workflow applications that do not require imaging functionality. The documents that are to be managed by the workflow system may already exist in a word processor, a spreadsheet, or a database application. Before considering the merits of image-enabled workflow it is important to understand this distinction.

If imaging is an important part of your organization's existing information systems or a component that you are considering to enhance the organization's workflow, you should be careful to evaluate the workflow system from the following perspectives.

First, consider the extent to which additional software and hardware will be required in order to facilitate an image-enabled workflow application. For example, some workflow systems can easily handle the ability to address and view an image, but many rely on an underlying image-processing application and hardware infrastructure. Simply being able to retrieve the image will not ensure adequate response time, network performance, or memory requirements. Since workflow acts as a layer of technology above the underlying data management applications (DBMS, imaging, text retrieval, etc.) you must first establish the viability of these technologies and then move on to the implementation of a workflow application.

Second, you should evaluate the approach that the workflow product takes towards the integration of imaging. Often ease of use will be a key consideration. Users typically favor an approach that minimizes the fragmentation of the interface environment. For example, if users are already comfortable with Windows, an approach that utilizes a Windows-based imaging system may appear to be ideal. But that is not always the case. Even a Windows-based imaging system may require that the user learn three separate metaphors for the desktop: one for the Windows environment, one for the workflow application, and yet a third for the imaging system.

Workflow can be designed, implemented, and used effectively without imaging.

The best approach is to provide a single metaphor for the desktop that will be consistent across each environment. These products will provide a metaphor for the desktop that includes icons and document objects. Images and word processing documents can be invoked and launched from the desktop for modification or viewing, which is fine as long as the user's environment remains the workflow desktop. But once an imaging application is launched, the user will need to work with the imaging application that manages the images as well.

Third, don't let imaging become a prerequisite for workflow. Workflow can be designed, implemented, and used effectively without imaging. Over time you may decide to augment your workflow systems with imaging. It is always wise to consider the long-term flexibility of the workflow application for new data types, such as images, but you may simply not be able to justify workflow if it is burdened with the investment required for fully functional imaging.

Document Management Facilities

The distinction between workflow and document management software is often blurred. In many cases, these two technologies can be used interchangeably to describe applications that manage the sharing of documents among multiple users and across networked environments. The distinction is that workflow provides a means to not only track versions, ownership, history, and retrieval, but also to productively manage the tasks that use the document. This enables the development of rules and sophisticated routing scenarios that are based upon more than the document alone.

At the same time, many workflow products do not offer a fully functional set of document management facilities, such as version control, content-based retrieval, security, and audit trails. Despite the apparent synergy that exists between workflow and document management, the two sets of functions are not always found in one product.

If document management is a key concern for your organization, you should consider the following.

First ask yourself, "What is the primary goal of the application?" If it is the tracking of documents, as opposed to the proactive management of document-based tasks, a workflow application may not be required. On the other hand, if document management is a secondary concern, a few basic facilities may be sufficient. These would include the ability to assign ownership to a document, set security levels for access privileges, and maintain version control across a distributed environment.

Second, document management may also require the ability to assign fixed fields, such as author, date, version, and index number, to document information. This can be done through the added facility of a document management system, but in many cases it can also be accomplished through an integrated database and development language. Be careful not to underestimate the power of a fully functional document manager or to overestimate your ability to build this level of functionality in a cost-effective manner. In almost all instances you are much better off choosing a workflow package with basic document management functions rather than attempting to develop your own.

Document Management and Workflow

Document management systems provide management controls over the filing and retrieval of electronic documents. "Rules" governing the management of individual documents are established within the document management system in much the same way that workflow systems require the establishment of rules relating to the management of work processes. Document management rules dictate such things as who has access to a document, who can edit a document, and how long the document will be maintained on the system.

Standard document management functionality includes the creation of a "document profile," which stores information relating to the attributes of a document. Generally the profile will contain information on the author, title, date, subject matter, and application. Most products allow the profile to be customized to meet the specific requirements of an application.

Another key feature provided by most document management systems is version control. This feature allows for the management of multiple versions of a document. Each time a revision is made to a document a new version is created. When the document is accessed, the most current version of the document is retrieved for viewing and editing. Earlier versions are retained and may be accessed but cannot be modified, thereby ensuring the integrity of the document and the audit trail provided.

Security and document integrity are essential elements of a document management system. Security and access control are provided at the document level. Individual documents can be made "public," shared within a limited workgroup, or they can be made "private" (restricting access to the author). Access rights can be assigned to groups of users, limiting their ability to view and modify documents.

Document management products also provide system administration features such as backup, archiving, report writing, and cost recovery tracking. Storage and retrieval services enable documents to be retrieved by attributes contained in the document profile. In addition, many document management products provide varying degrees of full-text search capabilities.

Document Routing and Workflow

While document management systems provide for the common storage and retrieval of documents on a network through the establishment of a shared document library, they typically do not provide proactive routing and controlled distribution of documents. For example, a member of a project team is given the responsibility of writing a report. Once complete, the report must be sent to another team member who is to comment, revise the report and forward the final document to the manager for approval. While the document management system allows for shared access to the report, it lacks the basic workflow functionality that would automatically route the document to the appropriate individuals for review and approval. As a result, manual methods of notification must be used, often causing extensive delays in the review process.

In such an example, a workflow system would control the process involved in routing the document to the appropriate individuals for review and would track the status of the document until its final approval and release. The workflow system however, would not control the management of the report document itself.

There is a natural synergy between workflow and document management that is often not exploited.

Most workflow products do not offer a fully functional set of document management facilities, such as version control, document profiling, and content-based retrieval, which are required to provide full life-cycle management. The inability to manage the individual document is a major shortcoming of many workflow offerings.

Combining Workflow with Document Management

There is a natural synergy between workflow and document management that is often not exploited. When the power of the two systems is combined, the result is full life-cycle management of both the work process and the documents involved in it. As vendors begin to capitalize on the benefits of uniting both technologies, it is becoming more difficult to identify products offered as either workflow or document management solutions. Generally workflow functionality provided in products intended primarily for document management is limited. Likewise many workflow offerings incorporate some document management functionality, but most do not offer fully functional document management capabilities.

Many document management systems vendors have expanded their capabilities by developing fully functional workflow modules as optional expansions to their existing document management systems offerings. Some vendors offer workflow modules as extensions to existing document management products. A few workflow products provide tightly integrated workflow and document management functionality.

Establishing Document Management Requirements for a Workflow System

It is important that planning for document management functionality be done up-front, because functionality required by the application will be a major consideration in evaluating and selecting a workflow product. When determining whether document management should be a component of a workflow system consider the following:

- What is the nature of the process being managed? Is it a repetitive transaction-based process containing large volumes of similar documents, or is it an ad hoc application that deals with diverse and often complex documents? Transaction-based applications generally do not require document management capabilities. If however, the process involves the preparation, editing, and revision of complex documents that must be tracked through a process, document management capabilities such as version control, document ownership, and security will be required.

- What is the orientation of the application? Is the primary objective the automation of a process, or control and organization of individual documents that are part of the process? If you want to provide both workflow and document management capabilities, you will need to establish priorities. If the main objective is to control documents on a network while providing a facility for routing documents, a document management product that provides basic workflow capability is adequate. But if the management of a complex work process is involved, you will need a more robust workflow product, containing full document management functionality or a workflow product that can be integrated with a document management system.

- Is there a method or system already in place for controlling documents? If not, how will documents included in the workflow system be managed? If there is a system in place, will it be sufficient to support the workflow system? A workflow system depends in large measure upon the ability to access documents needed to support work activities. If they are not adequately organized, they will not be accessible for use in the workflow application. This is an area often overlooked when planning a workflow system. Consistency in document naming and organization is critical in a workflow application, but very often access to documents in a networked environment is inadequate to support a workflow system. The ability to locate and retrieve documents needed to support a workflow application is essential. Document management functionality will provide the mechanism to organize files that support access to documents in the workflow system.

- How are documents retrieved? Do you need access by document attribute (title, subject, author, etc.) or is file name sufficient? Is full-text retrieval required? The ability to access documents by profile attributes and full text of the document can facilitate the workflow process, but fewer than 20% of available workflow products support full-text retrieval.

Office Automation Integration

Workflow applications are, by definition, integrated with other office automation tools such as word processing, E-mail, spreadsheets, and databases. Because these tools are diverse, and they often must work with separate collections of data, the workflow system must provide facilities to integrate each application. This may be accomplished using an operating system function such as OLE/DDE, through launching of the individual OA applications, or with the user environment (i.e., Windows). The specific method best suited for your workflow application will depend on the sophistication of your users and the integrity required of the integrated data sets. The integrity issue is the most difficult to address.

Integrity refers to the ability of the workflow environment to ensure that data modification across several data sets is synchronized in such a way as to avoid the processing of incomplete or unallowable transactions. For example, a customer support representative cannot submit an error-resolution request to engineering until the customer support supervisor signs off on it and the customer's account is checked to verify that their maintenance agreement is current. This may involve three data sets: the error-resolution-request document, the fixed data required for the supervisor's sign-off, and the database used to store financial information about the customer's account. All three must be coordinated in order to process the transaction. The workflow application must, therefore, access, retrieve, and potentially update one or more data sets.

Business Process Automation

Because workflow is as much a discipline as it is a technology, you would expect workflow vendors to provide a methodology for business process redesign and workflow simulation, not a technology alone. In all but a few cases you would be mistaken. Objective workflow methods are not prevalent. This obligates you to work with the vendor to define the problem as well as the potential solutions, although a few vendors have developed their own proprietary methods for workflow analysis.

Many organizations looking for an objective assessment of the potential solution believe that they are far better off obtaining an education and an open methodology that will work with any workflow technology. An interesting precedent in this regard is the decision of the Association for Information and Image Management (AIIM) to divorce its business analysis from the implementation of a solution. AIIM, the largest association in the imaging industry, needed to evaluate new technologies and re-engineering of their business processes. In issuing an RFP, AIIM decided to prohibit vendors who performed the business analysis from actually bidding on the solution. This ensured the objectivity of the analysis and recommended technologies. In addition, it keeps open the possibility of a no-go decision, which would, not be in the best interest of most technology vendors.

Off-the-Shelf Functionality versus Customization

Workflow products provide a range of capability from off-the-shelf user-oriented products to highly customized products that require a considerable amount of design and programming prior to installation.

The trend in the workflow market has been toward products that provide a combination of off-the-shelf and customizable functionality. As is so often the case, however, the best of both worlds is not always a union without compromise. Because most products have evolved from one of these two extremes, most products tend to favor one approach over the other. The key to identifying the product best suited to your needs is to establish how much customization your users and applications will require within a given period of time.

Transaction-based and object-oriented workflow products are designed to manage high volume, complex tasks that are repetitive in nature. For that reason transaction-based workflow products usually provide high-level job definition tools and usually require a substantial amount of customization. These products focus on performance and throughput rather than ease of use in job definition. The development of transaction-based products that offer graphical user interfaces and program libraries for end users is a definite trend.

Products requiring a high level of customization are unsuitable for the ad hoc workflow environment as ad hoc workflow products are highly individualized. Focus of evaluation for ad hoc workflow should be placed on ease of use and the effectiveness of job definition tools and the desktop environment. Products that emulate familiar desktop environments, such as the use of folders and other desktop icons, are easier to use than those requiring a high level of training and customization. The ability of end users to create and modify workflow applications on the fly is essential in an ad hoc workflow environment. Job definition tools should be graphically oriented, requiring a minimum amount of training to use. Workflow job definitions should be easy to develop and modify, thereby minimizing reliance on programmers and developers.

Some products provide a well-defined interface and object definition tools that enable end users to create workflow applications without programming. These products enable the end user to define and modify the workflow process in this way. The problem with this approach may turn out to be that the object parameters and supported rules are not robust enough to handle all of the possible contingencies. Workflow applications that contain complex procedures and rules will require customization beyond that provided by most icon-driven front-ends. Products that focus on complex workflow applications provide advance workflow definition tools and 4GL scripting that supplement basic off-the-shelf functionality. These provide limited user-definable workflow capability which can be enhanced through the use of advanced programming tools and APIs.

While many of the user-definable tools on the market are appealing, choosing one of these for a highly complex transaction-oriented application often sacrifices performance for end user ease of development. This could have a detrimental effect on the implementation and long-term success of the workflow application. Organizations considering a workflow application such as this should expect a high level of customization and should look for products that provide the highest performance available, while recognizing the need for ongoing technical support.

Scalability

The vast majority of workflow users will have to consider the migration of workflow across the enterprise. Few workflow applications will remain in a small workgroup for long. They quickly migrate into multi-department and enterprise-wide applications supporting larger numbers of users. In this scenario, the off-the-shelf package, which may be ideal for a small workgroup, will soon be outgrown as demands of users begin to diversify and create the requirements for specialized components that are not supported by the existing product. At the other extreme, it is unlikely that you can justify spending up-front time customizing a workflow application for small ad hoc applications involving a limited number of users.

Since few organizations initiate workflow on an enterprise-wide level, the ability to expand workflow applications is an important element to consider when evaluating a workflow product. You should be especially careful to select workflow products that provide open access to the documents and indices that have been used in the workflow application. This will enable you to migrate while preserving the investment in your data and access paths. It is also important to consider the compatibility of the workflow system with standardized user environments, such as Windows, and adherence to image standards, such as TIFF and CCITT compression. If the workflow product is also integrated with a standard DBMS, you will be able to preserve much of the structured information relating to the documents and data used in the workflow.

The vast majority of workflow users will have to consider the migration of workflow across the enterprise.

Functionality of Job Definition Tools

Job definition tools are the means in which a workflow application is written. The functionality of these tools varies significantly from product to product. Ad hoc workflow products usually contain graphical tools that allow end users to develop their own workflow applications, often on the fly. These tools are useful for the development of simple ad hoc workflow applications but are limited in their ability to handle more complex applications requiring a high level of integration. At the other end of the spectrum are the highly transaction-oriented workflow products, which contain a combination of graphical tools and 4GL scripting languages primarily for use by programmers for the development of customized applications. They are often difficult to use, requiring a high level of programming experience. Also, they are not easily modified, thereby requiring ongoing programming support for the maintenance of the system.

Workflow: In Search of Standards

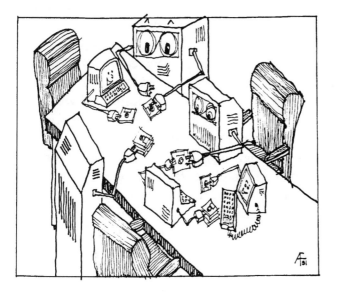

Nowhere is the need for standards greater than in the emerging workflow market. This chapter discusses the progress being made in developing workflow standards, through the efforts of the Workflow Management Coalition (WfMC) and the specific areas where standardization is needed.

Without the benefit of standards to govern the technology, workflow has evolved in many directions at once. Terminology, interface metaphors, methods, application integration, and implementation techniques vary from product to product. In some ways, this provides differentiation and market positioning advantages for vendors. On the other hand, it also discourages investment in products that may be quickly outdated and obsolete. With traditional information systems, that is always a consideration, but when investing in workflow products much more than just the information base is at stake—the process knowledge itself may be lost if it is locked up in proprietary repositories.

At such an incipient stage of the market it makes sense to consider carefully the stifling effect that an overly zealous approach to standardization may have on the infant workflow market.

Developments such as the formation of the Workflow Management Coalition (see the following section entitled *The Workflow Reference Model*) evidence the concern of not just users, but also the vendors, who are apparently realizing that the lack of process interoperability and standards severely impedes the enterprise application of workflow, slowing the market and stifling potential growth.

At the same time, any standard cuts both ways, slowing progress while better securing investment. At such an incipient stage of the market it makes sense to consider carefully the stifling effect that an overly zealous approach to standardization may have on the infant workflow market. But progress needs to made on this front if workflow is to be taken seriously as an enterprise tool.

If standards are to assist in the near-term implementation of multi-platform/process/ product solutions and still allow for the continued evolution of workflow there are two issues to address. First, the standard must provide a common workflow API set for multiple workflows as a process is passed from one product to another. This will enable an organization to use several workflow products while maintaining the continuity of a business process across workgroup boundaries. Second, the standard should enable the use of generalized metaphors, vocabulary, and tool sets for the definition of a workflow and a business process. This will create a workflow metaphor (think of spreadsheets or word processors) that allows developers and end users to buy into the concept of workflow through a set of familiar components that are predictable across workflow products.

Since workflow is a process technology, it will have to work with the range of applications involved in a process, such as E-mail, document management, groupware, and DBMSs. At this level, standardization is already underway in many areas that will have to consider workflow. For example VIM, MHS, or MAPI mail standards; open systems standards for file transport; document standards such as ODMA, Shamrock, OpenDoc, and OLE for document management applications; database standards such as SQL or ODBC; and an array of networking and communications standards.

If standards are to be developed at the API process level they should be sufficiently open to allow for some latitude in the business processes to which they are applied.

If standards are to be developed at the API process level they should be sufficiently open to allow for some latitude in the business processes to which they are applied. For instance, methodologies associated with those standards should also be open, allowing users to apply them with discretion and creativity. Business process redesign is, after all, a generalized science. It is not a cult that demands unyielding commitment.

Specific areas where we would probably see the greatest payback for immediate workflow standards include:

- Terminology and Definition[1]—A basic working vocabulary of workflow would go a long way to help vendors and end users speak the same language and measure the requirements of workflow applications against the functionality of workflow solutions on a common scale.

---

[1] Such a vocabulary is presented in this text. Many of the terms included have been presented as part of the industry standard/glossary published by the Workflow Management Coalition (WfMC).

- Interface Metaphors—Establishing a set of key interface metaphors for icons, folders, process flow, drill down, expansion, and depiction of workflow would assist in overcoming many of the conceptual barriers to workflow.
- Business Process Parameter Definition—A basic set of parameters that describes a task or an activity, which could be imported into any workflow product and would allow process definition to be independent of the workflow product.
- API set for the integration of standard office automation applications and basic sharing of tasks among workflow products by passing standard business process parameters.

These standards will establish a foundation for workflow interoperability. That, in turn, would lead to a more uniform market and an acceleration in the acceptance of workflow as a well-defined tool set with a common metaphor.

The problem, however, is simple. Innovation is causing workflow technology to move faster than the standards process. That leaves either a small cadre of vendors clinging to an outdated standard or it narrows the alternatives that users have to choose from in deciding on a workflow model.

The Workflow Reference Model: The First Step Toward Workflow Standards

Fortunately the balance between innovation and standardization is being sought, and may be at hand, through the efforts of a relatively new coalition of vendors and users. The Workflow Management Coalition, formed in October of 1993, has achieved substantial momentum, with more than eighty members who represent most workflow organizations. It is the only body creating standards for the workflow industry. (By contrast, in document management, for example, there are at least five competing [collaborating?] bodies—ODMA, OpenDoc, EDM/Shamrock, CORBA, ODA.)

The work of the coalition is based on the premise (a reasonable one so far) that all workflow systems contain a number of generic components that interact in a variety of ways. To achieve interoperability between workflow products, a standardized set of interfaces and data interchange formats is necessary. The Workflow Reference Model, developed by The Workflow Management Coalition, provides a working framework for the implementation of these standards. But, because workflow represents the intersection of many technologies and processes, the reference model contains a complex set of components, each of which requires standardization.

Five principal areas of standardization must be achieved for true enterprise workflow. These are as follows:

- The tools a developer uses to define a workflow,
- The workflow run-time module or server,
- The user interface,
- The links to other applications and services such as E-mail,
- The tool that passes work among different workflow products,
- The tools used to monitor a process.

Here is a brief overview of the five workflow components covered by The Workflow Reference Model and the specific standards that must be developed for each.

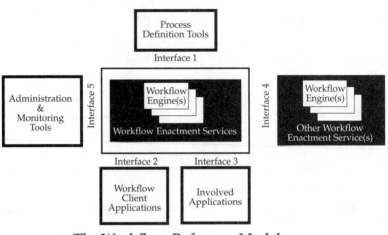

The Workflow Reference Model

Process Definition Tools and Interface

A variety of tools may be used to analyze, model, and describe a business process. The workflow model is not concerned with the particular nature of such tools, and currently each is in a form specialized for the particular workflow management software for which it was designed. One of the interfaces proposed by The Workflow Management Coalition enables more flexibility in this area.

This interface, termed the *process definition import/export interface*, would provide a common interchange format for the following types of information:

- Process start and termination conditions
- Identification of activities within the process, including associated tools and case data
- Identification of case data types and access paths
- Definition of transition conditions and flow rules
- Information for resource allocation decisions

Since there are other industry groups working in related areas such as process modeling and CASE tools, the Workflow Management Coalition intends to work with these groups to advance the definition of interchange formats with other non-workflow tools.

Workflow Enactment Service

The *workflow enactment service* provides the run-time environment in which one or more business processes are executed. This may involve more than one actual workflow server. For example, consider a business process that starts in Purchasing using FileNet's WorkFlo, then passes through Manufacturing, which uses UES's KiShell, then goes on to Sales, which uses Wang's Open/Workflow. In this case, a single process must be handed off from one workflow to another while continuity and process integrity is maintained.

The enactment service is distinct from the application and end user tools which are used to process items of work. This distinction provides the opportunity for a wide range of industry standard or application-specific tools to be integrated to provide a complete workflow application. This integration takes two forms: the tool invocation interface and the worklist handler interface. The *tool invocation interface* enables the workflow engine to activate directly a specific tool to undertake a particular activity. (This typically would be server-based and require no user action, for example, to invoke an E-mail application or pass data to a mainframe system.) The *worklist handler interface* through which the workflow engine interacts with a separate worklist handler responsible for organizing work on behalf of a particular user. Both are discussed in greater detail in the following sections.

Tool Invocation Interface

In addition to sharing information among themselves, there is also a requirement for workflow systems to deal with a range of third-party tools; for example, invoking an E-mail service such as X400, a fax service, document management services, or existing user applications.

There is significant value in the development of standards for the integration of such tools by building "tool agents" which will provide the interface the ability to invoke applications. These tool agents would also include a set of APIs which will allow other developers to build workflow-enabled applications. These applications can be invoked directly from the workflow engine, providing a high degree of integration and transparency to the end user. Coordination with other standards, such as ODMA and OMG, is essential in this area.

Worklist Handler Client and Interface

The *worklist handler* is like an electronic In-box. It is the software component that presents the end user with work items and the data related to them. The worklist may also invoke application tools associated with the work. The user takes action on the work, then passes the case back to the workflow enactment service. The worklist handler may be supplied as part of a workflow management product or may be a third-party product, such as E-mail, or a custom interface built in PowerBuilder, or Visual Basic, or another specific application.

The worklist handler provides all the functions needed for connecting the interface to the workflow engine and for obtaining and processing items of work. Effectively, this makes the workflow processing completely invisible to the end user. A worklist handler allows an enterprise to use any number of workflow products with a single consistent user interface.

Workflow Product Interoperability

Perhaps one of the most important objectives of the Workflow Management Coalition is to define standards that allow passing of work items among workflow systems produced by different vendors.

This objective can, however, be a double-edged sword. Workflow products are diverse in nature, ranging from those used for ad hoc routing to those aimed at highly regularized production processes. Each product has its own particular strengths. A single standard could, if zealously applied, polarize the market, forcing vendors to either choose the standard or exploit the strengths of their product by choosing a proprietary, and incompatible, approach. This could force workflow vendors to choose between providing a strong product focused on the needs of its customers and giving up those strengths just to provide interoperability at the level of the lowest common denominator. In its drive for interoperability standards, the Workflow

Interoperability can work at a number of levels from the systems with the ability merely to pass off tasks to workflow systems with complete interchange of process definition, case data and a common look and feel.

Management Coalition is determined not to stifle innovation.

Interoperability can work at a number of levels from the systems with the ability merely to pass off tasks to workflow systems with complete interchange of process definition, case data, and a common look and feel. However, the greatest level of integration is unlikely to be available generally as it relies on various developers using common approaches deep in their products—where healthy innovation is rife.

Workflow Status and Monitoring Application Interface

Finally, a common interface standard that allows one vendor's status-monitoring application to work with another engine will give workflow-enabled organizations the ability to track and report on an enterprise workflow application. A standard in this area means providing a complete view of work flowing through the organization regardless of which workflow system it is in. In addition, this standard would allow workflow users to choose the reporting and monitoring tool best suited to their needs, whether it be a statistical application or basic business graphics.

Summary of the Proposed Standards of the Workflow Management Coalitions Reference Model

- **Process Definition Import/Export**—A command set of process definition parameters that are produced by the tools through which developers or end users define a workflow.

- **Process Interoperability**—Standards that would allow multiple workflow products to co-exist and share the responsibility of managing a workflow.
- **Tool Invocation**—Standards for the initiation of an external tool, such as E-mail, document retrieval, or a user-written application.
- **Workflow Client**—Standards for the means of presenting work to be done to a process or an end user.
- **Status and Management**—Standards for collecting and storing certain key audit-based data about a workflow. These can later be used to identify the efficiency of a workflow.

In summary, standards will allow multiple products to work together and share workflow objects, much in the same way that SQL databases work. User demands for workflow standards are likely to be the driving force behind workflow development efforts and business buying decisions in the long-term.

Document Management Standards

The variety of standards, existing and proposed, that are attempting to define document interoperability is perhaps the biggest contributor to the evolution towards compound documents; and also the greatest liability. Since compound documents reflect a business process, they must be able to follow that process throughout the enterprise—even outside of the enterprise, to suppliers and business partners. Here is a list of the most popular Document Management standards.

SGML

SGML (Standard Generalized Markup Language) is an ISO (International Standards Organization) standard that has become the hottest topic in the document management arena today. SGML offers the ability to identify document sections in such a way that a document can be retrieved or, in the case of workflow, routed based on these structured items across a variety of SGML-compliant document platforms.

One of the problems with SGML, however, is that it does not provide a facility for associating foreign applications with the information objects it defines. In other words, the SGML document has no application intelligence. It is simply a means of identifying the structural components of a document. Workflow could provide this missing link.

ODA

ODA (Open Document Architecture) is similar to SGML in that it provides for structural identification of document components. But ODA goes further by defining a standard meta format for reconstructing document types across any platform or application. ODA-compliant applications need not be concerned with creating translators for specific document applications, instead they will support the ODA standard to achieve interoperability. The ODA effort faces the single greatest problem with establishing any common document architecture, vendor cooperation.

OLE

Although it is limited to Windows applications at the present time, Microsoft's OLE may turn into a universal client metaphor for compound documents. Among other things, OLE provides the ability to place information objects created in one OLE application into any other OLE application. In this way, an end user can invoke an information object directly from any OLE-compatible application.

CORBA

A competing standard, CORBA (Common Object Request Broker Architecture an OMG [Object Management Group] standard), backed by Apple, Borland, and WordPerfect, among others, provides a high-level container within which any variety of a compliant application's information types can be embedded. This offers a truly application-independent approach, at least to the extent of those applications supporting CORBA.

Shamrock

Shamrock is the code name for standards proposed by a set of industry vendors, lead by IBM and Saros, which includes over thirty member organizations. The group announced it would establish a set of APIs that would allow multiple document applications to share objects in a common repository structure. These are low-level standards that are best described as middleware between the document manager and the operating system. The first of these standards is ELS (Enterprise Library Services). The group is barely out of its embryonic phase but does have strong vendor support and funding.

ODMA

ODMA (Open Document Management API [Application Programming Interface]) recently announced (July 11, 1994) a set of standard APIs for document management integration with other applications, including workflow. These standards would allow desktop applications to work with compliant document managers with little or no integration required on the part of the end user. The group, which includes such vendors as WordPerfect, Soft Soutions, PC DOCS, and Documentum is actively moving ahead with the introduction of standards. Their fast pace may be a defining factor in their success.

DEN

DEN (Document Enabled Networking) will reside as "middleware" between the operating system software on network servers and the application software on clients. DEN will enable consistent access to documents located anywhere in workgroup or enterprise systems. Using this framework, application providers will be able to build applications on top of DEN that will take full advantage of its powerful document management capabilities. DEN's object-oriented architecture and integration of automatic conversion capabilities will allow organizations to protect investments in documents and maintain freedom of choice in the application programs used to create new documents.

OpenDoc

OpenDoc shifts documents from an application-centric approach to a process-orientation that uses application components, or appletts. These can be integrated seamlessly to create personalized application environments across multiple platforms. Since compound documents consist of multiple information objects (images, text, graphics, spreadsheets, etc.) the OpenDoc standard is capable of working with any combination of objects and applications within a single presentation, much like CORBA. OpenDoc was for some time considered to be direct competition to OLE, but is now supporting OLE 2.0 in an effort to become a comprehensive environment for compound document management. Supporting organizations are Apple, WordPerfect, IBM, Lotus, and Adobe.

The Future of Workflow

There will also be significant development in the assimilation of workflow into operating systems.

As workflow continues to infiltrate the enterprise, it will undergo an even greater metamorphisis, eventually transforming into a fundamental building block of information systems. Before that transformation can take place workflow must have greater ability to function as an analytical tool and a tighter integration with desktop environments.

Most noticeably lacking in current generation workflow products is the ability to simulate workflow procedures prior to implementation. At the present time there is little demand for simulation. The reason is that simulation requires a set of benchmarks and precise measures in order to provide reliable results. Although these measures may exist in factory environments, the critical mass of workflow has not yet been met to provide these for the office. Vendors will begin offering simulation tools within the next two years as business process redesign becomes more of a science. The process will not be unlike the flurry of educational activity generated by industrial engineering methods. The equivalent of work-process discipline is surely just arround the corner.

The most dramatic shift will occur over the next decade as workflow becomes the enabling technology for achieving the objective of business process automation.

There will also be significant development in the assimilation of workflow into operating systems. Ultimately leading to workflow's inclusion in future object-oriented Business Operating Systems. Desktop integration will be especially critical as workflow becomes a standard option for office automation environments and applications. This will create an increased awareness and demand for workflow, but it may also create problems as workflow developers begin to push the limits of low-end workflow. Without standards, such as those discussed earlier this could become a severe obstacle for the industry. It is likely that user demands for workflow standards, however, will be the driving force behind many development efforts as well as buying decisions.

A Revolution: The Final Chapter

The most dramatic shift will occur over the next decade as workflow becomes the enabling technology for achieving the objective of business process automation. As workflow brings together the splintered applications across the enterprise it will become the conduit that enables us to finally deliver on the promise of a paperless office. Ironically, when workflow reaches that point it will have become the most ubiquitous of all technologies— being nothing more than another icon on the desktop. These icons, or agents, will perform the myriad tasks of today's office worker. Each one will be the equivalent of an individual, a work group, or even an entire organization—complete with all of their process knowledge. These intelligent agents will automatically route documents to co-workers based on rules that represent the processes that govern the enterprise, accumulated through the business process analysis, simulations, and heuristics gathered over many years. These simple icons, which mask enormous complexity, will become the second most valuable asset of any enterprise.

What about the first most valuable asset of the enterprise, the people? Although it may be difficult to see beyond the current swath of employees being cut from the corporate heartland, the value of the worker is expected to increase more than at any other time in the history of mankind. It will be the final outcome of a revolution that began in factories.

The industral revolution was not born of the need to free workers from the menial; the early nineteenth century is littered with images of sweatshops and assembly lines. Craftsman who had built their knowledge and skill on the many generations of experience before them, gave way to specialization and mechanization. The craftsmen were transformed into cogs, and the cogs turned faster and faster with each decade. This new model of manufacturing demanded conformity, interchangeability, and rigid discipline. Ultimately, Fredrick Winslow Taylor's time-motion studies led to further segregation of work functions into myriad, precisely tuned components. The worker was demeaned and nearly dehumanized.

That legacy followed us through the factory, the office, every conference room and hallway of the enterprise—until unimaginable increases in computer power and affordability made it possible to re-invent the craftsmanship by making the tools and the knowledge needed to innovate readily available throughout the industrialized world.

We have traveled two hundred years and come full circle. As masters of more tools, knowledge, and experience than could have been wielded by 10,000 workers two centuries ago, today's knowledge workers are indeed the pinnacle of craftsmanship.

In tomorrow's enterprise, the knowledge worker will be freed to release creative energy that will result in an era of enormous innovation and discovery, fullfilling the potential and promise of the mind.

Workflow is the foundation for such far-fetched concepts to become reality. Ultimately workflow will unify the enterprise and make it impossible to separate the information systems from the business systems, developers from the users, blue collar from white collar. The diversity that leads to divisiveness will give way to the collaboration that breeds prosperity. Then we can say the revolution is over.

Afterword

Opportunity opens doors, but success requires walking through them. Workflow is an opportunity to change organizational information infrastructures to better support the business objectives of the organization, but it is not a guarantee of efficient processes, nor is it the only component of sound business systems. Workflow only provides the tools for change: analysis, compression, and automation of business cycles. By now that definition should be evident to the reader, as should the process by which the definition is put to use. What remains is the resolve to begin with a workflow effort, albeit even the smallest of starts. Doing that requires that you consider at least the following tenets:

- Evaluate your technology investment in parallel with your organizational analysis. One should not wait on the other. Consider the symbiosis of business process and information systems—each requires a clear understanding and appreciation of the other. It is rare, if at all possible, to understand the effect of a technology on a business process or to fully unravel the complexity of a business process without bringing technology and process together. That is the purpose and value of incremental change, such as that proposed by Stair Step.

- Who is the sponsor? Does the sponsor have the authority to commit funds to the project without further approval? Will he or she be committed for the full duration of the payback cycle?
- Establish the education level of your organization. Users, executives, developers, and even vendors must all communicate using the same vernacular and understanding of the process.
- Never assume that you know your infrastructure. Unless you have an up-to-date System Schematic, you don't. Define an infrastructure that meets the need of the process model. But do not let infrastructure define the process. If you do your application will become another island joined by paper bridges. Process boundaries and infrastructure boundaries must correspond precisely.
- Allow yourself the liberty of creating leverage through an incremental approach to change. Efforts to effect sudden, total change often result in total blunders. Success, even on a moderate scale, builds a track record of justification.
- Involve information systems management in the process. That is not an option, only a matter of timing. Involve them at the outset and your workflow systems will have the highest likelihood of integration with each other.

- Measure the process with the workflow tool.
  Metrics provide the fuel for change and
  minimize its risk. Consider both the impact
  of transfer time and task time in measuring
  your process efficiency. Use these measures
  as the foundation for your re-engineering
  efforts.

Who do you bank on in such a diverse market?
First, look for vendors who are making significant
headway in the areas of graphical desktops, integration
with standard DBMS technologies, use of object-
oriented techniques, intelligent document objects, and
cross-platform support. Second, don't invest in
workflow technology that replaces your desktop;
instead look for technology that works with the existing
metaphors that you have established. Existing E-mail,
workgroup, and desktop systems offer the best
leverage point for introducing workflow to your users.
Finally, and most important of all, remember that
ultimately workflow is more that just a technology; it
is an overall environment and approach to uniting
and automating business processes. As tantalizing as
it may be to daydream about obliterating existing
information systems, the reality is not so luxurious for
most of us. Workflow offers a means of measuring
and automating business processes incrementally.
That will provide a solid foundation and, more
important, a precise education on the benefits and
obstacles of organizational change.

Workflow is an education that will continue in
organizations as a perpetual exercise of re-invention.
As a colleague is fond of saying, "Everything must
justify its existence every day." That is a high calling
for executives, managers, and information systems
professionals, who have been trained to stake out a
direction, a business process, or a market, and be
unswervingly faithful to it. Although strategies should

not change at whim, the tactics and the methods by which we respond to external factors and achieve strategic objectives must be as adaptive in the office as they are in the factory. Empowering employees to cause change and take action is not an ideal state for a few users but instead must become a standard for all users. In that environment, workflow becomes an imperative for the adaptive enterprise.

Glossary of Workflow Terms[1]

Ad hoc
A workflow process model that is of temporary usefulness. As with any process, rules are part of an ad hoc process, but they cannot be well established and reused. Ad hoc rules tend to be based on judgment and heuristics that cannot be easily distilled into an automated process.

Agent
An automated task or process which can, after it is defined, be performed without human intervention.

AND-Join
When two or more parallel executing activities converge into a single common thread of control.

AND-Split
When a single thread of control splits into two or more threads in order to execute activities in parallel.

Application Data
Data that is application specific and not accessible by the workflow management system.

Asynchronous Teams
A group of individuals who can communicate through a workflow process, in which they are all involved, without the requirement of simultaneous human interaction. The process becomes the intermediary for communication between the individuals and roles that make up an asynchronous team.

---

[1] Includes terms from the Workflow Management Coalition *Glossary: A Workflow Managment Coalition Specification.*

Audit Trail
A historical record of the state transitions of a workflow process instance from start to completion or termination.

Business Process
A kind of process that supports and/or is relevant to business organizational structure and policy for the purpose of achieving business objectives. This includes both manual processes and/or workflow processes.

Case
A single instance of a process model, also referred to as an instance or an instantiation.

Completion
The definite conclusion of a task, process, or event as defined by the rules that govern a process.

Control Console
A visual display of workflow volumetrics for a given set of work objects, roles, individuals, tasks, or any other process component. Control Consoles are typically used to identify the status of a work in process, including measures of productivity and throughput.

Document
A collection of information objects authored for presentation and consumption by an end user. The key differentiator of a document over other data types is the fact that somone has invested intellectual capital to author the information.

Divergence
Also called a Split, the point at which multiple routes to multiple tasks are created from a single route. Contrast with *Rendezvous*.

Duration
The time required to complete one or more tasks.

Early Binding
The process of incorporating any changes to rules or objects at the time of a workflow process creation. In an early binding workflow system, any changes made to objects stored in a central library will not be reflected in the use of those same objects within workflows that reference the library. Contrast with **Late Binding**.

Event
Another term often used to refer to a task or a route. Many events have no duration. For example, upon the occurrance of the event "PO Authorization," the Purchase order is routed to accounts payable. An event is often synonymous with a trigger or an *initiation*.

Folder
A logical organization of information objects as a single work object. A folder is routed as one object, although it may contain many documents and data types. Folders may also be represented through the use of file cabinet, file drawer, or other file management metaphors.

Initiation
The event that triggers a task.

Iteration
A workflow process activity cycle involving the repetitive execution of workflow process activity(s) until a condition is met.

Knowledge-based

Workflow that incorporates into the workflow process model heuristic information about ongoing processes through the use of an inference engine, artificial intelligence, or other means. Knowledge-based work-flow may automatically shorten processes by aquiring intelligence, leading to self-modification.

Late Binding

The process of re-incorporating any changes to rules or objects at the time of execution. In a Late Binding workflow system, any changes made to objects stored in a central library will always be reflected in the use of those same objects within workflows that reference the library. Contrast with *Early Binding*.

Manual Process Activity

The manual process steps that contributes toward the completion of a process.

Manual Process Definition

The component of a process definition that cannot be automated using a workflow management system.

Manual Process Execution

The duration in time when a human participant and/ or some non-computer means executes the manual process instance of a process instance.

Manual Process Instance

Represents an instance of a manual process definition which includes all manual or non-computerized activities of a process instance.

Notification

An action which sends a message to an individual, role, or agent. A notification may not necessarily trigger a task. Notification can be forward or backward. For example, forward notification may warn of an upcoming event requiring additional resources.

Object-oriented
An information sytem that provides for the communication of information objects and process rules as a single entity. Object-oriented systems are characterized by key differentiators such as inheritance and encapsulation, which allows easy creation and replication of workflow applications.

Object
A combination of rules, procedures, and information in a single entity.

Organizational Role
A synergistic collection of defined attributes, qualifications and/or skills that can be assumed and performed by a workflow participant for the purpose of achieving organizational objectives.

OR-Join
When two or more workflow process activities physically connect or converge to a single activity. In this case there is no synchronization of the threads of control from each of the two or more workflow process activities to the single activity.

OR-Split
When a single thread of control makes a decision upon which branch to take when encountered with multiple branch(es) to workflow process activities.

Parallel Routing
A segment of a workflow process instance where workflow process activity instances are executing in parallel and there are multiple threads of control.

Predecessor
A task or route prior to the current task, or route.

Priority
A process attribute which determines the sequencing of information objects through a workflow.

Procedure
A single logically interrelated set of tasks. Often used to refer to a subset of a *process model*. Procedures may be represented as single objects in some workflow systems.

Process
A coordinated (parallel and/or serial) set of process activities that are connected in order to achieve a common goal. A process activity may be a manual process activity and/or a workflow process activity.

Process Activity
A logical step or description of a piece of work that contributes toward the accomplishment of a process. A process activity may be a manual process activity and/or an automated workflow process activity.

Process Activity Instance
An instance of a process activity that is defined as part of a process instance. Such an instance may be a manual process activity instance and/or a workflow process activity instance.

Process Definition
A computerized representation or model of a process that defines both the manual process and the automatable workflow process.

Process Definition Mode
The time period when manual process and/or automated (workflow process) descriptions of a process are defined and/or modified electronically using a process definition tool.

Process Execution
The duration in time when manual process and workflow process execution takes place in support of a process.

Process Instance
Represents an instance of a process definition which includes the manual process and the automated (workflow process).

Process Independence
The ability to define and execute a process independent of the underlying technologies that enable it. Process independence allows for a process to be automated across what would otherwise become technology boundaries.

Process Model
The highest level of definition for a workflow process. Process models may consist of subprocesses called *procedures*, which are independent process models.

Process Role
A synergistic collection of workflow process activities that can be assumed and performed by a workflow participant for the purpose of achieving process objectives.

Queue time
The time a piece of information, ready for a task to be performed against it, waits for the next task in its process to be performed. Queue time must always follow transfer time.

Receivership
The ability to organize communications around the information requirements of the recipient instead of the sender.

Rendezvous
The point at which multiple routes from multiple tasks are rejoined to proceed as a single route to only one successor task. Contrast with *Divergence*.

Resolution Routes
The path an information object or a process takes in order to reach a conclusion. There are two types of resolution routes: one to n, which requires more than one task; and one and done, which requires only one task.

Role
A specific set of skills required to perform a given set of tasks.

Routing
The logical, defined transfer of information through a process and its associated tasks, based on specified rules. There are five possible routing architectures: Serial—each task has only one predecessor and only one successor; Parallel—a group of tasks have the same successor and the same predecessor task; Concurrent—same as parallel, but the tasks between predecessor and successor must begin and end at the same time; Conditional—a situation in which multiple routes may be followed based on a rule, procedure, or variable; Dependence—a route that is predicated on the completion of another task. In effect, every route is dependent on its immediate predecessor task being completed; however, a dependence routing scheme explicitly states this dependence on a task that may not be an immediate predecessor task.

Rule
A defined parameter that determines the action taken against an information object.

Script
The rules of a process expressed as a procedural definition of a workflow, typically in traditional coding format.

Sequential Routing
A segment of a workflow process instance where workflow process activity(s) are executed in sequence.

Splicing
The ability to combine ad hoc workflow with transaction-based workflow through the splicing of a single task in a transaction-based process model, and then inserting the ad hoc workflow route. Object-oriented systems are the only ones capable of this type of function.

Status
The measurement of a work in progress.

Sub Process Definition
A process that is called from another process or sub process that includes the manual process and the automated (workflow process) components of the process.

Successor
The task or route immediately following the current task or route.

Suspense
The halting of a process or an information object prior to completion. Usually suspense refers to a suspense queue, which is a holding area awaiting the initiation of a task.

Task

A finite set of actions which have a defined initiation and conclusion. For example, the approval of a purchase order by a manager is a task. Tasks have task time associated with them and are preceded and succeeded by transfer time.

Task queue

The logical storage area for an information object waiting for processing.

Task time

The time to complete a given task.

Tool

A workflow application that interfaces to or is invoked by the workflow management system via the workflow application programming interchange/ interface.

Transaction-based

Workflow that is used for repetitive processing of like transactions.

Transfer time

Elapsed time from the completion of one task to the start of the next task with sequence. Transfer time may include queue time if it is not measured separately.

Transition Condition

Criteria for moving, or state transitioning, from the current workflow process activity to the next workflow process activity(s) in a workflow process instance be it a manual process or a workflow process.

Transmit time

The time required to physically transmit an information object, ready for a task to be performed against it, to its immediate successor task queue.

Value Chain
A series of tasks or processes which ad value to an information object as it progresses from task to task. A value chain can be represented by organizational groupings as well as process groupings.

WAPI
The application programming interface/interchange for client workflow applications and tools in order to be able to interface to the Workflow Enactment System. WAPI is an acronym for **W**orkflow **A**pplication **P**rogramming **I**nterface/**I**nterchange.

Work
The process of adding value through the performance of one or more tasks.

Work-cell
A grouping of individuals or roles that work together on a common task or process. Most often this is represented by project teams. Unlike traditional organizational groupings, however, individuals can belong to multiple work-cells, whereas they only belong to a single department.

Work Item
Representation of work to be processed in the context of a workflow process activity in a workflow process instance.

Work Item Pool
A space that represents all accessible work items.

Work Object
A collection of items, which may also include rules, that is routed collectively through a workflow process.

Workflow
A proactive tool set for the analysis, compression, and automation of information-based business cycles.

Workflow Enactment Service

A software service that may consist of one or more workflow process engines in order to create, manage and execute workflow process instances. Client workflow applications/tools interface to this service via the workflow application programming interface (WAPI).

Workflow Interoperability

The ability for two or more workflow engines to communicate and interoperate in order to coordinate and execute workflow process instances across those engines.

Workflow Management Coalition

An international body of vendors and users, established in 1993, for the furtherance of standards and interoperability among workflow products.

Workflow Management System

A system that completely defines, manages and executes workflow processes through the execution of software whose order of execution is driven by a computer representation of the workflow process logic.

Workflow Participant

A resource which performs partial or in full the work represented by a workflow process activity instance.

Workflow Process

The computerized facilitation or automated component of a process.

Workflow Process Activity

The computer automation of a logical step that contributes toward the completion of a workflow process.

Workflow Process Activity Instance
An instance of a workflow process activity that is defined as part of a workflow process instance.

Workflow Process Control Data
Data that is managed by the Workflow Management System and/or a Workflow Engine.

Workflow Process Definition
The component of a process definition that can be automated using a workflow management system.

Workflow Process Engine
A software service or "engine" that provides part or all of the run time execution environment for a workflow process instance.

Workflow Process Execution
The duration in time when a workflow process instance is created and managed by a Workflow Management System based upon a workflow process definition.

Workflow Process Instance
Represents an instance of a workflow process definition which includes the automated aspects of a process instance.

Workflow Process Monitoring
The ability to track workflow process events during workflow process execution.

Workflow Process Relevant Data
Data that is used by a Workflow Management System to determine the state transition of a workflow process instance.

Worklist
A list of work items retrieved from a workflow management system.

Worklist Handler

A software component that manages and formulates a request to the workflow management system in order to obtain a list of work items.

Suggested Readings

Ackoff, Russell L. *The Democratic Corporation*. Oxford University Press, 1994.

Baker, Joel Arthur. *Paradigms: The Business of Discovering the Future*. Harper Collins, 1992.

Belasco, James A. Ph.D., *Teaching the Elephant to Dance*. Plume, 1991.

Davenport, Thomas. *Process Innovation*. HBS Press, 1993.

Davidow, William H. and Michael S. Malone. *The Virtual Corporation*. Harper Collins, 1992.

Denna, Eric L., J. Owen Cherrington, David P. Andros, and Anita Sawyer Hollander. *Event-Driven Business Solutions*. Business One Irwin, 1993.

Drucker, Peter F. *Managing for the Future*. Truman Tally Books (Penguin), 1992.

Garson, Barbara. *The Electronic Sweatshop*. Simon and Schuster, 1988.

Hammer, Michael and James Champy. *Reengineering the Corporation*. Harper Collins, 1993.

Hunt, Daniel V. *Reengineering*. Oliver Wright Publications, 1993.

Kuhn, Thomas S. *The Structure of Scientific Revolutions*. The University of Chicago Press, 1970.

McCluhen, Marshal. *Understanding Media*. Signet Books, 1964.

Morris, Daniel and Joel Brandow. *Re-engineering Your Business*. McGraw-Hill, 1993.

Naisbitt, John. *Reinventing the Corporation*. Warner Book, 1985.

Sashkin, Marshall and Kenneth J. Kiser. *Putting Total Quality Management to Work*. Berrett-Kohler, 1993.

Schor, Juliet B. *The Overworked American*. Harper Collins, 1991.

Schrage, Michael. *Shared Minds*. Random House, 1990.

Seidman, William L. and Steven L. Skancke. *Productivity: The American Advantage*. Simon and Schuster, 1989.

Taylor, Fredrick W. *The Principles of Scientific Management*. Harper Brothers, 1911.

Toffler, Alvin, *Future Shock*. Bantam Books, 1971.

Index